Diese

Questions and Answers books are available on the following subjects:

Automobile Brakes and Braking
Automobile Electrical Systems
Automobile Engines
Automobile Steering and Suspension
Automobile Transmission Systems
Car Body Care and Repair
Diesel Engines
Light Commercial Vehicles
Motor Cycles
Electricity
Electric Motors
Electric Wiring
Amateur Radio
Radio Repair
Radio and Television
Colour Television
Hi-Fi
Electronics
Integrated Circuits
Transistors
Brickwork and Blockwork
Carpentry and Joinery
Painting and Decorating
Plastering
Plumbing
Central Heating
Refrigeration
Electric Arc Welding
Gas Welding and Cutting
Pipework and Pipe Welding
Lathework
GRP Boat Construction
Steel Boat Construction
Wooden Boat Construction
Yacht and Boat Design

QUESTIONS & ANSWERS

Diesel Engines

J. N. Seale

revised by

John Hartley

Newnes Technical Books

The Butterworth Group

United Kingdom	**Butterworth & Co (Publishers) Ltd** London: 88 Kingsway, WC2B 6AB
Australia	**Butterworths Pty Ltd** Sydney: 586 Pacific Highway, Chatswood, NSW 2067 Also at Melbourne, Brisbane, Adelaide and Perth
Canada	**Butterworth & Co (Canada) Ltd** Toronto: 2265 Midland Avenue, Scarborough, Ontario, M1P 4S1
New Zealand	**Butterworths of New Zealand Ltd** Wellington: T & W Young Building, 77–85 Customhouse Quay, 1, CPO Box 472
South Africa	**Butterworth & Co (South Africa) (Pty) Ltd** Durban: 152–154 Gale Street
USA	**Butterworth (Publishers) Inc** Boston: 10 Tower Office Park, Woburn, Mass. 01801

First published 1964 by George Newnes Ltd
Reprinted 1966, 1969, 1971, 1975, 1976, 1977, 1979
Second edition 1980 by Newnes Technical Books

British Library Cataloguing in Publication Data

Seale, Jack Norman
Diesel engines. – 2nd ed. – (Questions & answers).
1. Diesel motor
I. Title II. Hartley, John III. Questions and answers on diesel engines IV. Series
621.43'6 TJ795 80–40262

ISBN 0-408-00474-6

Typeset by Butterworths Litho Preparation Department
Printed in England by Fakenham Press Ltd, Fakenham, Norfolk

CONTENTS

PREFACE

With the need to conserve energy, the diesel engine is becoming a more important source of power for vehicles as well as for industrial equipment. There is also an increasing trend towards the use of the diesel engine in cars.

This book explains the working principles of the diesel engine and includes examples of some modern engines. It has been extensively rewritten and a large number of the illustrations are new.

The accent is on automotive and car diesels. For example, one section is devoted to turbocharging, another to car diesels, and there is a completely new chapter on changes to suit the environment – the reduction of noise, smoke and fuel consumption. There is a substantial section on injection equipment and another new chapter covers auxiliary equipment. The aim throughout has been to present the information clearly, so that it will be of interest not only to those involved with diesels but also to those who are simply interested in them. Units used are SI (metric) and a conversion table is included.

I would like to thank those people and companies who supplied information and illustrations for this revised edition. Particular thanks are due to Lucas CAV, Bosch, Volkswagen AG and Cummins Engine Co. Ltd.

J.R.H.

1

PRINCIPLES OF OPERATION

What is meant by the term 'diesel' or 'oil engine'?

An internal-combustion engine which uses a heavy or gas-oil distillate as fuel, the fuel being fired in the cylinder by heat of compression. It is, therefore, also known as a compression-ignition engine.

How did the diesel engine originate?

In the engine invented by Dr. R. Diesel, which first appeared in 1893, the heavy oil fuel was injected into the engine cylinder mixed with a jet of compressed air. This *air-blast* injection principle is not used today.

In the engine invented by Ackroyd Stuart, two years before Dr. Diesel filed his first patent, the oil was injected into the engine cylinder without any compressed air, i.e. as a spray or mist of oil particles. A hot bulb or other external source of heat was used for ignition. This *airless* or *solid* injection principle is now used on the majority of modern diesel engines.

What types of fuel oil are used in diesel engines?

The principal constituent of the fuel used in diesels – called diesel fuel, DERV or gas oil – is 'gas oil', which is a medium-temperature distillate of crude oil. Diesel fuel is similar to kerosene, and its critical characteristic is that it must ignite easily when compressed and a certain temperature is reached. This is the opposite of the requirement for petrol which must not ignite spontaneously. In low-speed diesels fuel quality is less critical, and they can operate on a mixture of gas oil and the thicker residual oils.

Why is the diesel engine classed as an internal-combustion engine?

Both diesel and petrol engines convert the latent heat energy of the fuel into mechanical work. The heat in the fuel is released by combustion of the fuel in air. The diesel is classed as an internal combustion engine because the air is involved in the combustion process.

In the steam engine the working fluid does not take part in the combustion of the fuel, and so it is not an internal combustion engine.

What is the basic difference between a petrol engine and a diesel engine?

In the petrol engine, a mixture of air and petrol vapour is drawn into the cylinder during the suction (or induction) stroke of the piston and is ignited towards the end of the compression stroke by means of a sparking plug.

In the diesel engine, only air enters the cylinder during the suction stroke – it may enter by suction or by blowing in under pressure. The fuel oil, in the form of a fine spray or mist, is injected into the cylinder towards the end of the compression stroke and is spontaneously ignited by the temperature of the highly-compressed air.

What is the 'swept volume' or 'capacity' of a cylinder?

This is the space or volume swept out in the cylinder by the piston, and is equal to the product of the piston stroke and its cross-sectional area.

What is the 'clearance volume'?

This is the space or volume remaining in the cylinder and/or cylinder head where the air charge is compressed by the piston.

What is the 'compression ratio' of an engine?

The compression ratio of an engine is the ratio of the sum of the swept and clearance volumes in each cylinder to the clearance volume.

If, for example, the cylinder holds 64 cm^3 of air when the piston is at the bottom of its stroke, and there is only a 4 cm^3 space left when the piston is at the top of its stroke, the engine has a compression ratio of 16:1.

What are the usual compression ratios for petrol engines?

Between 7:1 and 10:1.

What are usual compression ratios for diesel engines?

Between 12:1 and 24:1. Generally speaking, compression ratios are higher in small high-speed diesel engines than in large low-speed diesel engines. The compression ratio also depends on the type of combustion chamber: open or swirl chamber type.

What are the two basic types of diesel engine?

The two-stroke (or two-stroke cycle) engine and the four-stroke (or four-stroke cycle) engine.

What is meant by a 'two-stroke' engine?

One working stroke occurs (in each cylinder) for every revolution of the engine crankshaft, i.e. every alternate stroke is a firing stroke.

What is meant by a 'four-stroke' engine?

One working stroke occurs (in each cylinder) for every two revolutions of the engine crankshaft, i.e. there are one firing or working stroke and three idling strokes for every two revolutions of the crankshaft.

What is the fundamental difference between two-stroke and four-stroke engines?

In the two-stroke engine, a separate pump or blower is required to recharge the cylinder with air, whilst in the four-stroke engine the working cylinder itself performs that duty.

Describe the cycle of operation in a typical two-stroke engine

When the piston is at the bottom of its stroke (*Fig. 1(A)*), it fully uncovers the air-inlet (scavenge) ports arranged circumferentially round the cylinder wall just above the top of the piston. Air from a blower enters the cylinder through these ports and scavenges the exhaust gases out through the open exhaust valve in the cylinder head.

As the piston moves upwards (*Fig. 1(B)*), it covers the air inlet ports, the exhaust valve closes and the piston compresses the fresh air charge to a fraction of its original volume, say one sixteenth, and to a pressure of up to 70 bars, depending upon the design of the engine.

Just before the piston reaches the top of its stroke (*Fig. 1(C)*), atomised fuel is sprayed into the cylinder. The high temperature of the compressed air ignites the fuel spray and the resulting pressure forces the piston downwards.

On the downward stroke of the piston (*Fig. 1(D)*), the exhaust valve opens and the cylinder is swept with clean scavenging air as the piston uncovers the inlet ports.

This entire combustion cycle is repeated for each revolution of the crankshaft.

Is there another arrangement for exhausting the burnt gases in a two-stroke engine instead of using exhaust valves in the cylinder head?

Yes. A number of engines employ exhaust ports circumferentially round the cylinder wall in a manner similar to the air-inlet ports.

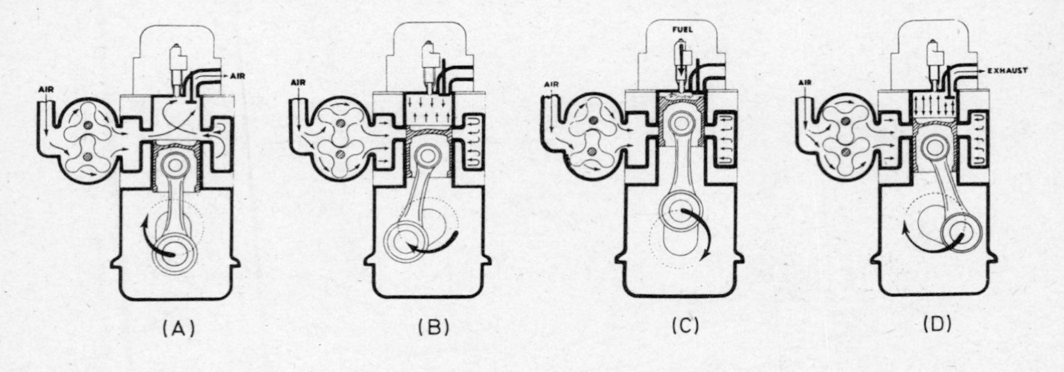

(A) *Piston at bottom dead centre:*
Scavenge ports arranged circumferentially round the cylinder wall, above the piston in its lowest position, admit pre-compressed air from a blower. The flow of air sweeps out the exhaust gases through the open exhaust valve.

(B) *The upstroke:*
As the piston moves upwards it covers the scavenge ports and the exhaust valve closes and the charge of fresh air is compressed to about 1/16th of its original volume. This happens on every upward stroke of the piston in a two-stroke cycle engine.

(C) *Piston at top dead centre:*
Shortly before the piston reaches its highest position atomised fuel is sprayed into the cylinder. The high temperature of the compressed air ignites the fuel spray and the resulting pressure forces the piston downwards.

(D) *Piston nearing bottom dead centre:*
On the downward stroke, the exhaust valve opens and the cylinder is swept with clean scavenging air as the piston uncovers the inlet ports. This entire combustion cycle is repeated for each revolution of the crankshaft.
(*General Motors*)

Fig. 1. The two-stroke cycle (one piston per cylinder)

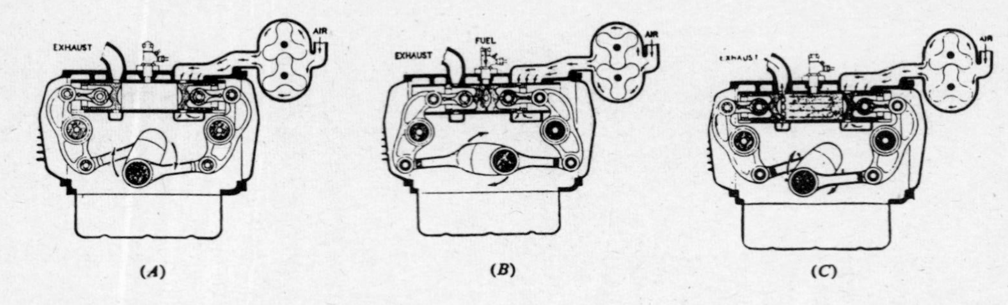

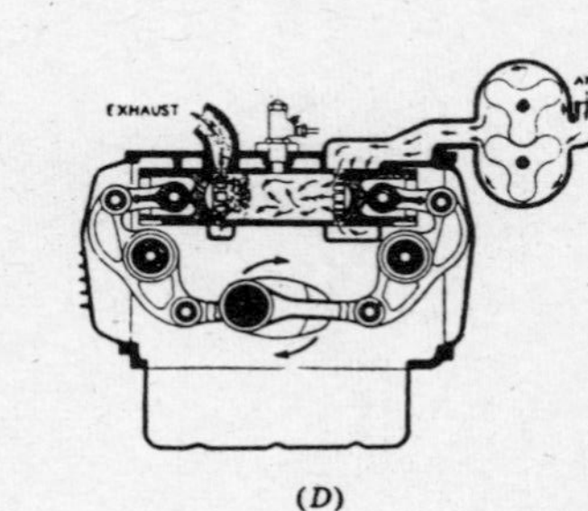

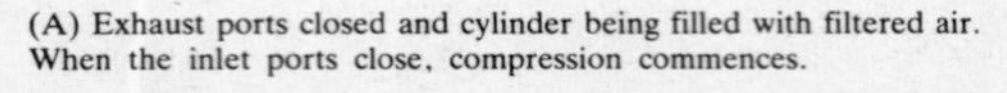

(A) Exhaust ports closed and cylinder being filled with filtered air. When the inlet ports close, compression commences.

(B) The air is compressed, thereby raising its temperature. As pistons near inner dead centre, fuel is injected into cylinder. This is ignited by the high temperature of the compressed air and the power stroke commences.

(C) Power stroke ends when exhaust ports are uncovered, as the gases rush out through the ports, causing a drop in pressure within cylinder. This is known as 'blow down'.

(D) Air ports are uncovered and 'blow down' ends. Fresh air charge supplied under pressure by blower scavenges cylinder until exhaust ports close, when cycle is repeated.

Fig. 2. The two-stroke cycle of a Commer horizontally-opposed-piston engine

How do horizontally-opposed-piston two-stroke engines operate?

In the opposed-piston two-stroke engine, such as the Commer TS3 or the Leyland L60, a pair of pistons oppose one another in the same cylinder. Their connecting rods may bear directly on a crankshaft, there being two crankshafts per engine, or may actuate rockers which transmit the power to the crankshaft.

While the motion of one piston controls the inlet ports in the cylinder wall, the other one controls the exhaust ports, and the fuel injector is mounted in the wall of the cylinder between the pistons. As the piston passes the inlet ports, so the blower forces air into the cylinder. Then, when the piston covers the ports, the air is compressed, thus raising its temperature, and near top dead centre fuel is injected and ignited. The combustion causes the pistons to be pushed away from the centre of the cylinder until eventually the exhaust ports are uncovered to allow the gases to be discharged. Soon afterwards, the inlet ports are uncovered again, and fresh air enters the cylinder, forcing the remaining exhaust gases out.

Among the advantages of this layout is the uni-directional flow, which can be arranged to spiral to give good mixing with the fuel. Also, the absence of gaskets and valves enables the engine to operate at high pressure. However, opposed-piston engines are very complicated and costly.

What is the cycle of operations in a four-stroke engine?

The sequence of operations can best be described under the four strokes of the piston: induction, compression, firing and exhaust.

Induction stroke (*Fig. 3(1)*): As the piston moves down the cylinder, due to the suction effect, it draws air into the cylinder through the open inlet valve. The exhaust valve remains closed during this stroke.

Compression stroke (*Fig. 3(2)*): The piston moving up the cylinder, with both inlet and exhaust valves closed, compresses the air to about one sixteenth of its volume and to a pressure of up to 70 bars, depending upon the design of the engine.

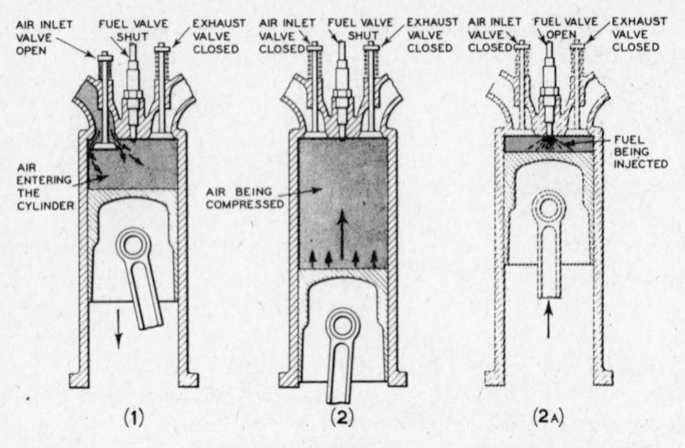

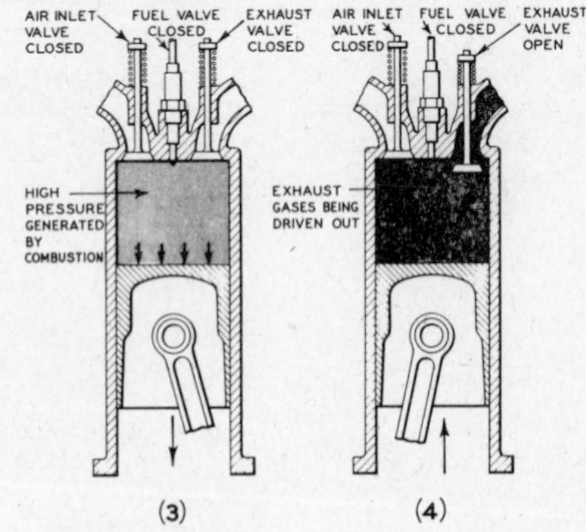

(1) *Induction stroke:*
The piston moving down the cylinder draws in a charge of air through the open air inlet valve, the fuel and exhaust valves being closed.

(2) *Compression stroke:*
All valves are closed and the piston moves up the cylinder compressing the air to about 1/16 of its original volume and thereby raising its temperature.

(2A) *Fuel injection and ignition:*
Just before the piston reaches the top of its stroke the fuel valve opens and a spray of fuel oil is injected into the cylinder, the high temperature of the compressed air being sufficient to ignite it.

(3) *Firing stroke:*
High pressure, generated by combustion, drives the piston downwards on its firing or working stroke.

(4) *Exhaust stroke:*
Near the end of the firing stroke, the exhaust valve opens and as the piston returns towards the cylinder head the burnt gases are driven out. At the top of the stroke the exhaust valve closes and the air inlet valve opens.

Fig. 3. The four-stroke cycle

Firing stroke (*Fig. 3(3)*): Just before the piston reaches the top of its stroke, atomised fuel is sprayed into the cylinder. The high temperature of the compressed air ignites the fuel spray and the resulting pressure forces the piston downwards on its firing or working stroke. Near the end of the firing stroke, the exhaust valve opens to allow the burnt gases to escape.

Exhaust stroke (*Fig. 3(4)*): In moving up the cylinder the piston drives out the remaining products of combustion through the open exhaust valve. Near the end of the exhaust stroke, the inlet valve opens, the exhaust valve closes and the cycle of operations is repeated for every two revolutions of the crankshaft.

On many engines, the exhaust valve does not close until just after the piston has passed the top dead centre position and is descending on its induction stroke.

How is the cylinder scavenged of burnt gases on two-stroke and four-stroke engines?

In two-stroke engines, air blown into the cylinder through ports at the end of the firing stroke increases the speed at which the exhaust gases are driven out through the open exhaust valves or ports. The scavenging air is provided by a scavenging pump or blower.

On four-stroke engines, the cylinder is scavenged of burnt gases initially by the piston moving up the cylinder on its exhaust stroke, and secondly by air entering the cylinder during the inlet- and exhaust-valve overlap period.

What is meant by 'valve overlap'?

To assist in removing the exhaust gases from the cylinder, the inlet and exhaust valves remain open together for a period of time. This is achieved by opening the inlet valve shortly before the piston reaches its top dead centre position (at the end of the exhaust stroke) and closing the exhaust valve shortly after the piston begins to descend the cylinder on its induction stroke.

What type of flow is used to scavenge the exhaust in two-strokes?

Various systems have been used, but probably the best-known on modern engines are the Schnuerle and uniflow systems. In the

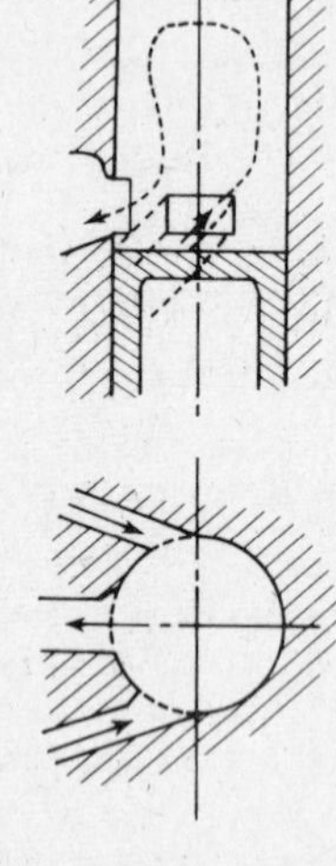

Fig. 4. Schnuerle loop-scavenging system for two-stroke engines

Schnuerle system (*Fig.4*) the air is directed upwards through inclined transfer ports to the rear of the cylinder, so that it sweeps across the roof of the cylinder head and down to the exhaust port.

What is the uniflow system?

In the uniflow system, shown in *Fig. 1,* the gases do not change direction in passing through the cylinder. Therefore, the inlet ports are at one end of the cylinder and the exhausts are at the other – this system is used mainly on opposed-piston two-strokes (*Fig. 2*).

How is the exhaust gas prevented from contaminating the fresh-air charge in a two-stroke engine where the air ports are either level with or higher than the exhaust ports?

By the use of an automatically-operated non-return scavenge valve in the inlet manifold. This valve opens to admit the blown air into the cylinder after the piston has uncovered the exhaust ports.

How does the Kadenacy system of scavenging two-stroke engines work?

The Kadenacy system utilises the energy in the exhaust system to provide a depression in the cylinder, the partial vacuum, which in turn causes scavenge air to flow into the cylinder behind the escaping exhaust gases. A blower may also be used with this system, but it is not an essential feature.

The system is based on the assumption that when the exhaust ports or valves are opened rapidly during the expansion stroke, there is within the first time interval of a few thousandths of a second, an urge or impulse in the gases to escape very rapidly from the cylinder, thus leaving a depression behind them. The fresh charge of air is allowed to enter the cylinder behind the exhaust gases by suitable timing of the admission valve or ports.

What is meant by the 'combustion chamber' in a diesel engine?

The combustion chamber is the space into which the atomised fuel spray is injected and ignited by the high temperature of the compressed air.

What are the three main types of combustion chamber for high-speed engines?

(1) Direct injection (DI) or open chamber.
(2) Swirl, or separate turbulent chamber.
(3) Feed, or pre-chamber.
(2) and (3) are termed 'indirect injection' (IDI) systems.

What is an open combustion chamber?

Usually, the combustion chamber is formed between the flat or slightly coned face of the cylinder head and a cavity machined centrally in the piston crown, the fuel injector being situated in the cylinder head and over the centre of the cylinder.

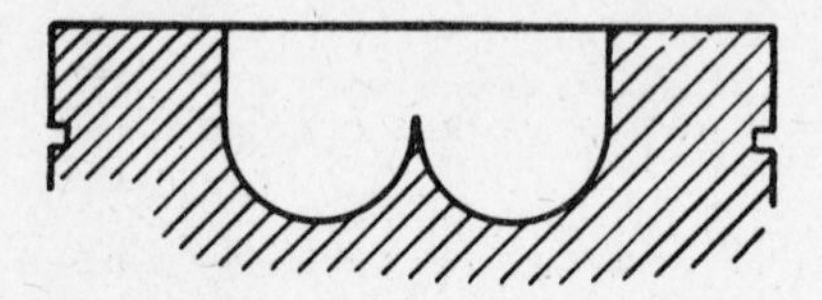

Fig. 5. Straight-sided toroidal chamber in piston crown

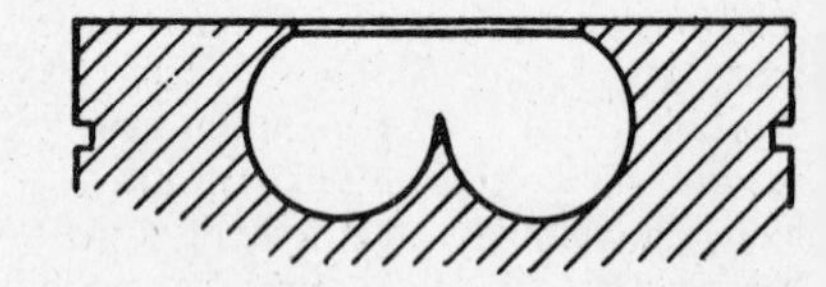

Fig. 6. Restricted-opening toroidal chamber in piston crown

The shape of the cavity in the piston crown may be either hemispherical, very nearly spherical, cylindrical, bowl-shaped or toroidal (see *Figs. 5* and *6*).

Can the open combustion chamber as applied to vehicle (high-speed) engines be termed a 'swirl chamber'?

Strictly speaking, yes. This is because it utilises an organised swirling movement of the air in the cylinder to enable the combustion reactions to keep pace with the engine revolutions.

To induce air swirl in the cylinder, the air is directed tangentially as it enters the cylinder by either shrouding the inlet

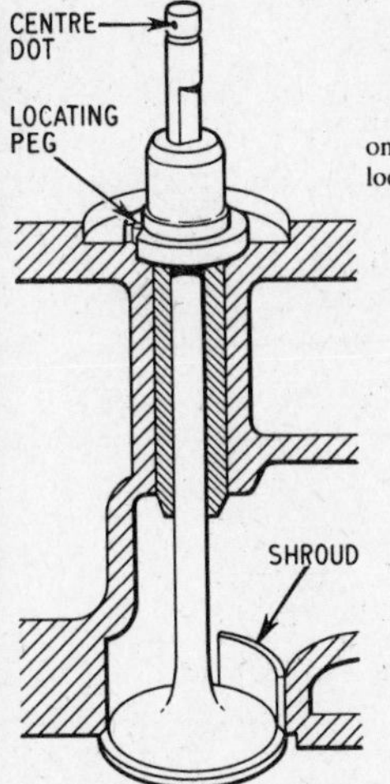

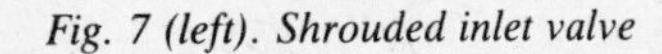

Fig. 7 (left). Shrouded inlet valve

To ensure correct location of the shroud in the cylinder head on this engine, the valve-stem dot should be adjacent to the locating peg for the valve thimble.

Fig. 8 (below). Directed inlet port

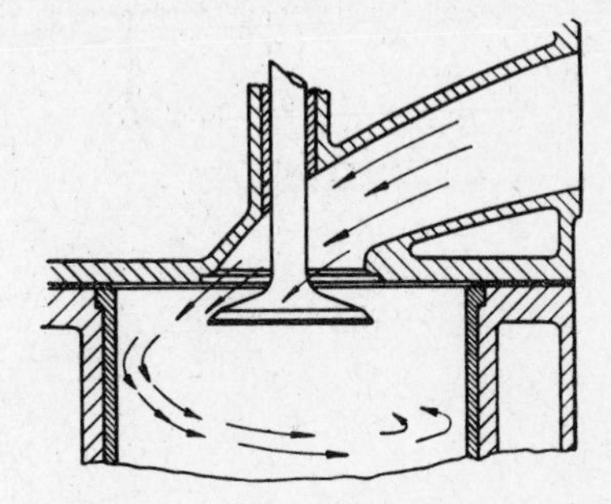

valve (*Fig. 7*) or disposing the inlet passage (*Fig. 8*), or, as in the case of two-stroke engines, by tangential ports.

What is the swirl combustion chamber used in high-speed engines?

This is a shaped cavity which is usually quite separate from the engine cylinder. The air displaced from the cylinder during the compression stroke is set into rapid rotational movement within the swirl chamber.

There are several varieties of swirl chamber, but probably the best known is the Comet developed by Sir Harry Ricardo.

In the Ricardo Comet Mk. III combustion chamber (*Fig. 9*), 50 per cent of the air volume is transferred into the swirl chamber during the compression stroke, and the balance, after deducting that in the piston-cylinder-head clearance, is contained in two shallow pan-shaped depressions machined tangentially to each other in the piston crown.

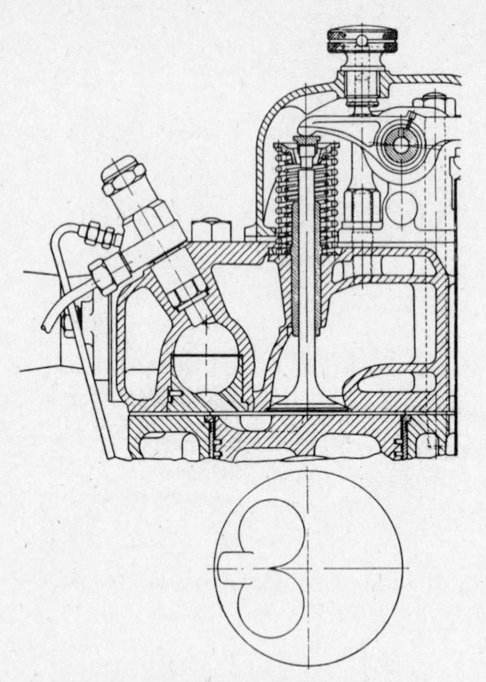

Fig. 9. Ricardo Comet Mk. III swirl-type combustion chamber

The air enters the swirl chamber from the cylinder through a throat or passage which is tangential to the swirl chamber to provide the desired rotary motion.

After partial ignition of the fuel in the chamber, the rush of gas from the throat is directed into the depressions in the piston crown and forms a pair of oppositely rotating vortices which bring the air contained in them into contact with the burning mixture issuing from the swirl chamber.

In the later Mk. V version of the Comet (*Fig. 10*), the side of the separate lower part of the swirl chamber to which the fuel is injected is made approximately in the form of a truncated cone.

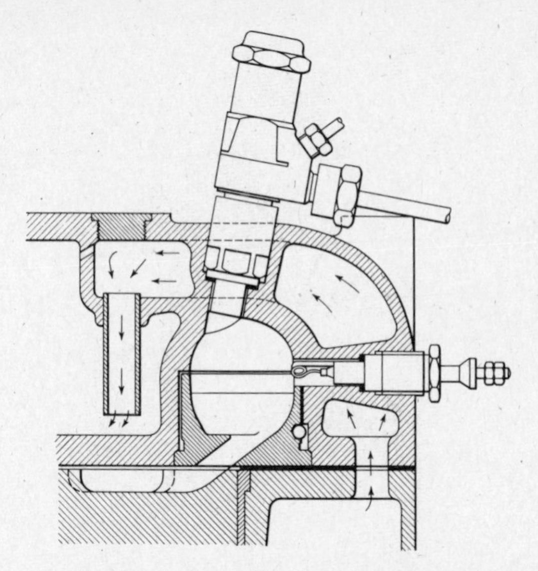

Fig. 10. Ricardo Comet Mk. V swirl-type combustion chamber

The other side retains its spherical shape. The upper half of the chamber, which is cast in the cylinder head, also remains hemispherical in shape.

What is the pre-combustion chamber for high-speed engines?

This form of combustion chamber, although not favoured in the UK, has proved popular in the rest of Europe and the US.

The chamber is divided into two parts. The larger part is a shallow, saucer-like cavity machined in the crown of the piston; the smaller, in the cylinder head, forms the pre-chamber (or antechamber) – Fig. 11. The pre-chamber, which is roughly pear-shaped, is connected to the main chamber through an orifice, termed the 'burner', placed at its smaller end. The fuel injector is fitted at the top, or larger, end of the pre-chamber.

Towards the end of the compression stroke, fuel spray is injected into the pre-chamber where it is thrown on to the inner side of the burner which, under operating conditions, becomes heated. The heated sides of the burner and the high temperature of the compressed air ignite the fuel and the resulting rise in

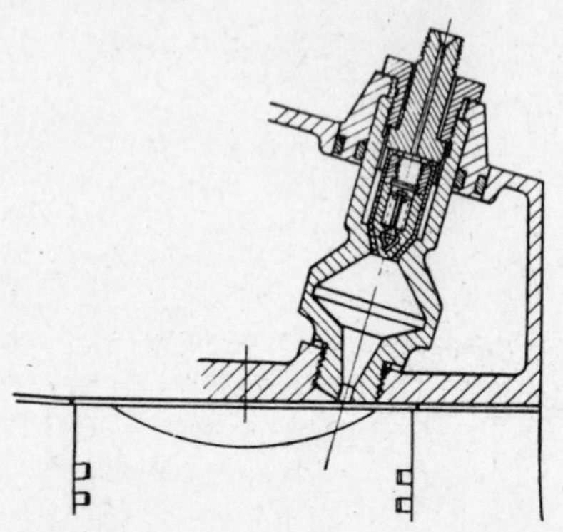

Fig. 11. Pre-combustion chamber of the Caterpillar Tractor Co.

pressure causes a rush of gas from the pre-chamber, driving out the fuel into the main chamber where it completes its combustion with the main volume of air.

In some types of burner, a number of small holes converge towards a single large hole which forms the final outlet to the main chamber. Other burners use a single hole tapering down to the actual orifice.

How is the fuel oil delivered to the engine cylinder?

By means of a fuel-injection pump delivering fuel oil under pressure to the injector on the cylinder. The nozzle valve or needle in the injector is raised off its seat at the appropriate time by the pressure of the fuel and this allows the fuel oil, in the form of a fine spray or mist, to enter the engine combustion chamber through the hole or holes in the end of the injector.

What are alternative names for fuel injectors?

Fuel injectors are also known as fuel valves, atomisers, nozzles and sprayers.

What is the 'injection pressure' of the atomised fuel?

A wide range of injection pressures is used in diesel engines, the choice of pressure for a particular engine being governed by its design and application and the size of nozzle hole or holes.

The pressure depends principally on the rate of injection required, which in turn depends on the size of the cylinder and the amount of turbulence in the air movement. Thus, in large engines the trend is towards the use of very high pressures of 1000 bar, with pressures of 600–750 bar in automotive direct injection engines. In IDI engines, much lower pressures of about 200—500 bar are used.

What is meant by 'supercharging', or 'pressure charging'?

Supercharging (pressure charging) is a system of introducing air into the engine cylinder at a pressure higher than atmospheric, thereby increasing the quantity consumed, and thus allowing for a proportionate increase in liquid fuel to be burnt to provide an increase in the power output of the engine.

How is the supply of compressed air obtained in supercharging systems?

The supply of compressed air is obtained in a variety of ways, e.g.:

(1) Engine-driven compressor or blower.
(2) Ram effect.
(3) Supercharging.
(4) Exhaust gas turbochargers.
(5) Exhaust gas pulse supercharging (Comprex).

What are the main types of compressor or blower used for supercharging?

(1) Lobe or eccentric-vane blower (*Fig. 12*).
(2) Centrifugal blower.

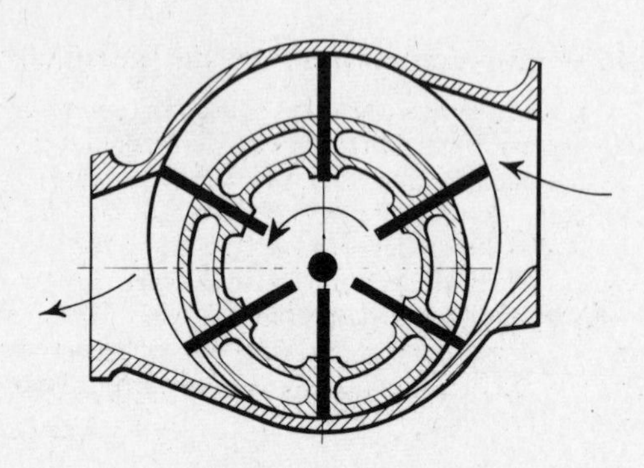

Fig. 12. Eccentric-vane-type blower

What is 'ram effect'?

If the lengths of the inlet pipes are set so that they induce resonant harmonic air oscillations, the kinetic energy provides a ramming effect which can increase volumetric efficiency. However, the effect is relatively small compared with other forms of supercharging, and best results are obtained at one speed only.

What is 'aftercharging', or the 'topping up' method of supercharging?

During the latter part of the suction stroke on a four-stroke engine, compressed air is admitted to the cylinder (which is already filled with induction air at atmospheric pressure) at sufficient pressure to provide the desired degree of supercharge. The compressed air is normally admitted to the cylinder through a specially designed inlet valve.

What is the most widely used and efficient system of supercharging?

By using the engine exhaust gases to drive a turbine which is directly-coupled to a rotary air compressor or blower supplying compressed air to the engine intake (see *Figs. 13* and *14*).

Why is turbocharging used widely?

The advantage of turbocharging is that the energy used – that in the exhaust – would otherwise be wasted, and so no extra power is consumed by the turbocharger. Engine-driven superchargers can consume a lot of power, especially at high speeds.

How much can a turbocharger increase the power output?

The power increase depends on the application and on the breadth of the speed range required. In most automotive applications, where a wide speed range with good low-speed torque is required, the power is usually 25–50% greater than for a naturally-aspirated engine. However, high-pressure-ratio turbocharged engines are now being introduced which develop 60–80% more power than naturally-aspirated units, while an increase of 250% is obtained in some military engines.

What is the exhaust gas pulsing system of supercharging?

This is the Comprex system developed by Brown Boveri, in which the pulses in the exhaust gases are used to pressurise the inlet air, also in pulses. The Comprex unit is a rotor with radial vanes on a spindle, the rotor being enclosed in a housing. At one

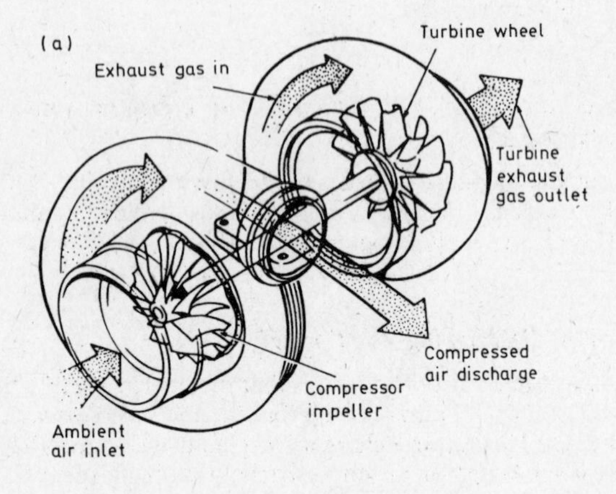

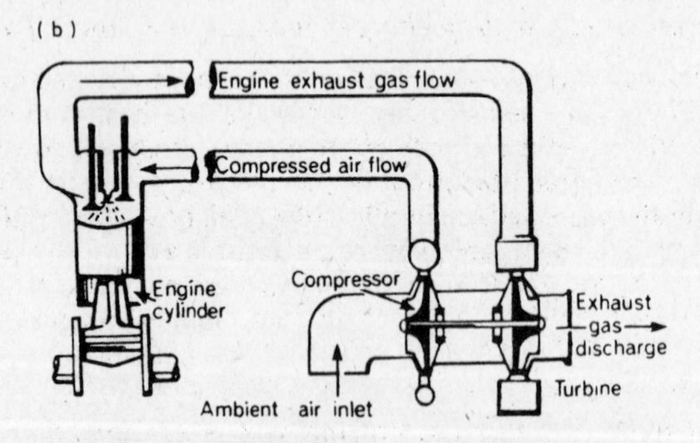

Fig. 13. Schematic drawings showing (a) inlet and exhaust gas flow through the turbocharger itself, and (b) when connected to an engine

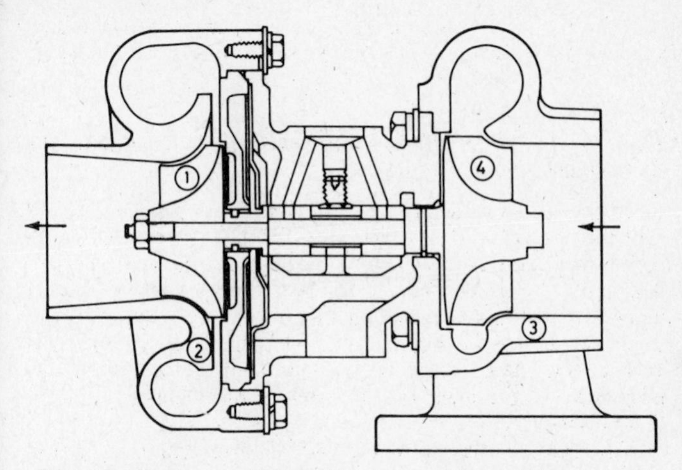

Exhaust gases from the engine drive the turbine wheel, 1, and then are discharged through the housing, 2, to the exhaust. Air entering the compressor housing, 3, is compressed by the compressor wheel, 4, and discharged through the volute to the exhaust manifold.

Fig. 14. An automotive supercharger (see Fig. 51)

end there is an inlet port connected to the exhaust manifold, and adjacent to it is a discharge port which extends to the exhaust pipe. At the other end of the housing is another pair of ports. The inlet is connected to the air filter, and the outlet to the inlet manifold of the engine. The rotor is driven by the crankshaft, but is not timed.

Thus, on each exhaust stroke, exhaust gases are forced through the pipe and into the housing, the gases passing along a passage between a pair of vanes. Since the housing is filled with air the exhaust gases compress this air against the end of the housing. Then, as the rotor rotates, so the space between the vanes moves opposite the outlet port, and the air is forced up through the inlet manifold to the engine. The rotor continues to rotate, so the exhaust gases strike the end of the housing, and are reflected back so that they pass out through the outlet port to the exhaust system.

As the exhaust gases are discharged they leave a partial vacuum or depression in the housing, so more fresh air enters ready for the next cycle.

What is the basic difference in performance between the turbocharger and Comprex?

Since the turbocharger relies on a centrifugal compressor and centrifugal turbine, the exhaust gases must overcome the inertia of the rotating members before they can accelerate them and deliver extra power. Therefore, the turbocharger is best at higher speeds, and cannot match the engine over the complete speed range unless compound turbocharging is used, but this is very costly. Also, there is a slight delay before the boost increases, so care is needed in matching the fuelling system to the turbocharger.

With Comprex there is no inertia to be overcome, so response is virtually instantaneous. In addition, the system tends to give more low-speed torque than a turbocharger, so it is particularly suitable for commercial vehicles. However, it is a relatively expensive unit.

2

ENGINE TYPES AND APPLICATION

What are the main applications of diesel engines, in terms of engine speed?

(1) *Low-speed engines* (*95–500 rev/min*):
Large marine propulsion engines (direct-coupled or geared reduction drive to propeller).

(2) *Medium-speed engines* (*500–1000 rev/min*):
Marine propulsion units for trawlers, tugs, coasters and ferries (geared-reduction or diesel-electric drive).

Stationary work. Ideal for continuous-running applications, such as land and marine generating sets, pumping systems (water, sewage, gas distribution lines, oil pipelines). Also as automatic stand-by generating sets in power stations and industrial establishments.

Rail traction (shunting locomotives, rail-cars and mainline diesel-electric locomotives).

(3) *High-speed engines* (*over 1000 rev/min*):
Rail traction (as above).

Commercial vehicles, coaches, buses, private cars, taxicabs, and agricultural tractors.

Industrial applications including electric power generation.

What are the usual types of diesel engine?

In addition to engines being either normally-aspirated or supercharged, engine types can be classified by their cylinder arrangements. The most general designs are as follows:

(1) Engines with their cylinder axes vertical or horizontal, or inclined as in V engines (two- and four-stroke types).

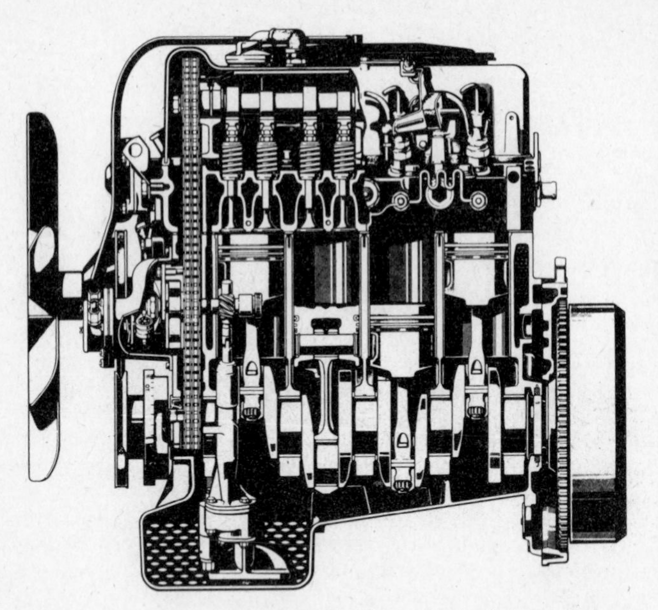

Fig. 15. The Mercedes-Benz OM616 four-cylinder 2.4-litre engine develops 53 kW at 4400 rev/min

(2) Horizontally- and vertically-opposed piston engines (two-stroke type).

(3) Large marine propulsion engines – single-acting crosshead and trunk engines (two- and four-stroke types), and double-acting crosshead engines (two-stroke type). Single-and double-acting, two-stroke, crosshead engines may also be of the opposed-piston type.

How do vertical (including V) and horizontal engines of medium power and speed compare in choice for stationary work?

The medium-speed horizontal engine which, for a number of years, held the market for pumping, stand-by generating sets and

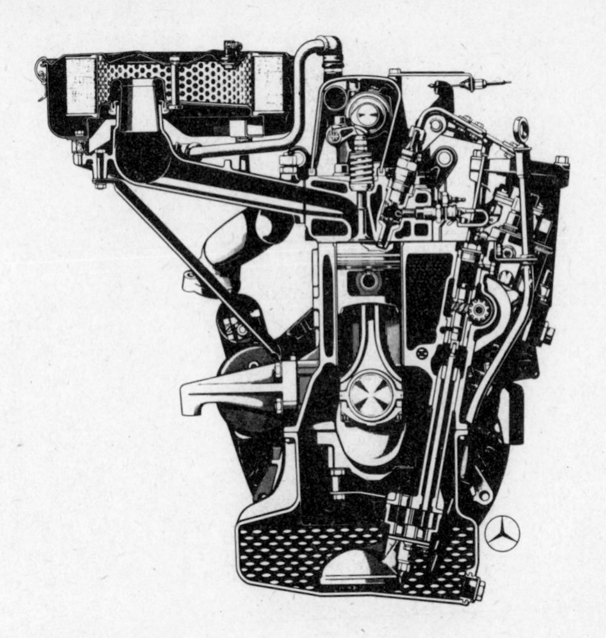

(*Fig. 15 continued*)

belt-drive applications, has now been largely superseded by the vertical engine. This is mainly due to the lower weight per unit power and the smaller floor-space requirements of the vertical engine.

What advantages has the diesel engine over the petrol engine for transport vehicles?

The chief advantages are that diesel oil is cheaper than petrol, and the fuel consumed in a given distance is considerably less than for petrol.

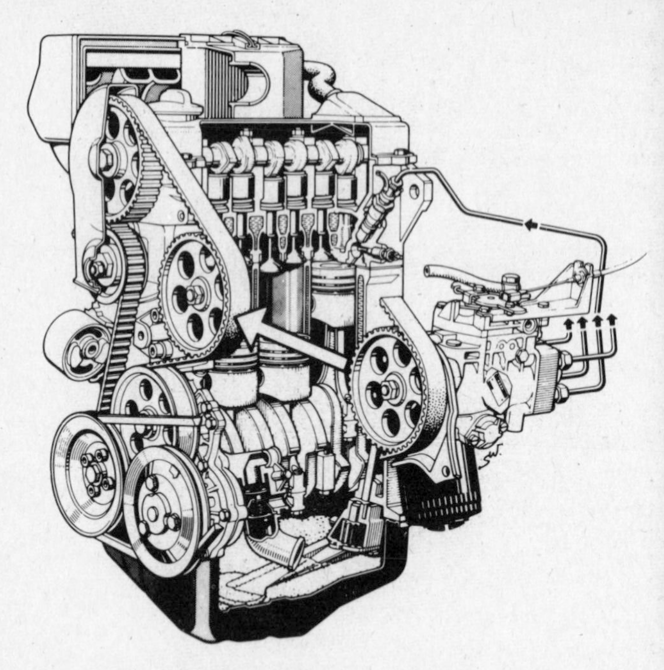

Fig. 16. The VW Golf diesel of only 1.5 litres is based on the petrol engine

What is the main disadvantage?

The cost of a diesel engine is more than that of a petrol engine of the same power, although this higher initial outlay is offset by lower running costs.

Can high-speed automotive and rail-traction engines be adapted for industrial applications?

With a few alterations, yes.

How do two-stroke and four-stroke engines compare in field of choice?

Both types of engine are in equal competition in the high- and medium-speed field ranges, but the two-stroke-cycle engine is almost in universal use for continuous low-speed applications, such as marine propulsion.

What type of two-stroke engine employs exhaust valve(s) in the cylinder head instead of exhaust ports arranged circumferentially round the cylinder wall?

The uniflow-scavenged, two-stroke engine employing one piston per cylinder.

How are diesel engines rated in terms of load?

On diesel engines it is necessary to fix the maximum load which the engine will carry for a given time. These loads are the 'rated' loads of the engine and are normally given by the three following forms of rating:

(1) The twelve-hour rating, which is the 'basic full load'.

(2) The one-hour rating, or 'overload', which is 10 per cent greater than the twelve-hour rating.

(3) The continuous or twenty-four-hour rating, which is 10 per cent *below* the twelve-hour rating.

Ratings (1) and (2) refer to the continuous period of time for which the engine will carry the given load. The continuous rating (3) is that which the engine will carry day and night until mechanical wear demands an overhaul.

What is a 'peak' load?

This is quoted by some engine builders, and it refers to the load which the engine can carry for a few minutes under emergency conditions. The usual practice, however, is to fit a stop on the fuel-injection pumps to prevent them from delivering enough fuel to exceed the one-hour rating.

In what form is the 'rating' of an engine usually given?

In kilowatts (kW) at revolutions per minute (rev/min) of the engine crankshaft.

What is the 'brake' power of an engine?

The brake power of an engine is the useful power developed at the engine crankshaft after overcoming the frictional resistance of the engine itself. It is known as 'brake' power because the power can be absorbed and measured on a dynamometer – the dynamometer is termed a brake because it absorbs and dissipates the energy developed by the engine.

What governs the power output of an engine?

The total power of any engine is dependent upon its design, cylinder bore, piston stroke, number of cylinders, rated speed and its required application.

How does the power output of a naturally-aspirated diesel compare with that of a petrol engine?

The maximum power output of a naturally-aspirated diesel engine is invariably less than that of an equivalent petrol engine. Smaller, car-type diesels have higher specific outputs than larger ones, and these are up to about 20–23 kW/litre. A typical petrol engine for a saloon car has a specific output of 30–35 kW/litre.

Why is the specific power output of a diesel lower than that of a petrol engine?

The power output of the petrol engine is limited by the volumetric efficiency. As the speed rises near peak power, so the volumetric efficiency reduces until eventually there is inadequate mixture being drawn into the engine to increase the power output. Thus, by increasing the volumetric efficiency, the power output can be increased. In the diesel, the same amount of air is drawn in at all speeds, the power output being increased with more fuel being injected. As soon as more fuel is injected than can be burnt in the air, combustion is incomplete, and a lot of black smoke is emitted from the exhaust. Therefore, the power is limited to that which can be obtained without significant amounts of smoke being generated.

How does the turbocharger overcome this problem?

Since the turbocharger pressurises the air there is more air in the cylinder with which the fuel can burn. In theory, the turbocharged engine can develop an enormous amount of power, but in practice it is limited by the thermal stresses in the engine and the need to maintain a wide spread of power.

What are typical power outputs of current engines?

Outputs of naturally-aspirated diesels range from about 10 to 24 kW/litre, in automotive form. Industrial engines are usually rated at 10–12 kW/litre. A typical modern industrial engine is the Perkins 4.2032 Squish Lip engine (see page 96), which develops 44 kW at 2600 rev/min from a displacement of 3.3 litres, so the specific output is 12 kW/litre. Typical of engines used in trucks are the Perkins 6.354 (5.8 litres) and the Fiat 8.1 litre diesel. These develop 93 and 126 kW respectively, equivalent to 15.5–16 kW/litre, and peak power is developed at about 2600 rev/min.

Of the car diesels, the Mercedes-Benz 2.4 litre four-cylinder diesel is typical with an output of 54 kW at 4400 rev/min (22 kW/litre), while the Volkswagen developing 37 kW at 5000 rev/min (25 kW/litre) is typical of the smaller, new highly-rated diesels.

What are typical power outputs of turbocharged diesels?

Turbocharging is normally used to increase power output by 25–30% for engines to be used in trucks, but the trend is to higher specific outputs. Generally, therefore, turbocharged truck engines have specific outputs of about 20 kW/litre. A typical example is the Perkins T6.354 which develops 116 kW at 2600 rev/min (20 kW/litre). Rolls-Royce has exploited turbocharging to give much higher specific outputs on its large CV-8 and CV-12 diesels for industrial and military use. For commercial use, the 17.4 litre V-8 is rated at 410 kW at 2100 rev/min (23.6 kW/litre), but in military form, the output is increased to 600 kW (34 kW/litre). The trend is towards this sort of output in the future.

What are the general features of modern diesels?

Although there are two main types of diesel – DI and IDI – these engines have many common features. Nowadays, virtually all engines have separate cylinder blocks and heads, these generally being iron castings. The detachable sump may be either an aluminium casting or pressed steel. To keep noise emissions as low as practicable the cylinder block is a robust casting, usually having one bearing housing between each pair of cylinders – that is, five main bearings on a four-cylinder unit, and seven on a six – and the block normally extends well down past the axis of the crankshaft.

Cylinder liners, dry or wet, are used on heavy-duty diesels. Dry liners, which are either a press fit or a sliding fit in the block, are preferred in most cases since this results in a stiffer cylinder block. However, wet liners are often used on very large diesels of over 10 litres displacement.

Liners are usually made of cast iron, while the pistons are of low-expansion aluminium alloy. The valves are installed vertically in the cylinder head, and are normally actuated by push-rods and rockers from the camshaft. An important feature of a diesel is the timing drive since this must drive the fuel injection pump as well as the camshaft, and so it is subject to high loads. In addition, it must maintain the timing of injection very accurately. Therefore to drive the camshaft, injection pump, oil pump and compressor/exhauster, it is normal to use gears and these are generally helical. On smaller engines, however, there is a trend towards the use of rubber toothed-belt timing drives.

What are the features of the Volkswagen 1.5 litre car diesel?

The Volkswagen 1.5 litre diesel has a bore and a stroke of 80 × 76.5 mm and it develops 37 kW at 5000 rev/min. Maximum torque is 79 N m (59 lbf ft) at 3000 rev/min. This unit is based on the VW petrol engine, using the same basic cylinder block, crankshaft, and connecting rods. The valve gear is also similar, as is the aluminium cylinder head. Therefore, this unit is nearer a petrol engine in concept than most diesels. It is an IDI engine, the combustion chamber being based on the Ricardo Comet Mark V. As is normal on this type of engine, the lower portion of the pre-combustion chamber is formed in a detachable insert of heat-resistant alloy. The compression ratio is very high at 23.5:1 and this helps with cold starting.

An interesting feature of the VW diesel is that it has a toothed-belt timing drive and a rotary-distributor-type injection pump. The valves are actuated by the camshaft operating through inverted bucket-type tappets and the induction manifold is longer than is common, its length being set to give some ram effect.

What are the features of the Mercedes-Benz OM616 diesel engine?

This engine is similar in many respects to a Mercedes-Benz petrol engine. With a bore and stroke of 91 × 92.4 mm, to give a displacement of 2.4 litres, its maximum power output is 54 kW at 4400 rev/min, and maximum torque 137 N m (101 lbf ft) at 2400 rev/min. The compression ratio is 21:1. Mercedes-Benz engines have the Daimler-Benz pre-chamber system (page 15) which, at the expense of some efficiency, gives very low combustion noise for a diesel.

The deep cast iron cylinder block has five main bearings and dry liners, and chain drive to the single ohc which drives the valves through finger-rockers. The pre-combustion chamber is inserted in the cylinder head at a slight angle from the vertical, and the cavity in the piston is deeper than in the Ricardo system.

. . . *and the Leyland TL12 diesel engine?*

The Leyland TL12, a typical heavy-duty automotive diesel in the 10–14 litre class, is a six-cylinder in-line unit with a deep cast iron cylinder block and massive seven-bearing crankshaft. To simplify manufacture, the counterweights are bolted to the crankshaft webs. The crankpins are drilled to reduce weight and provide a mass of oil near the bearing, the ends being sealed by cups. There are dry cylinder liners, and the camshaft is housed alongside the lower ends of the water jacket.

On the nose of the crankshaft is a gear that drives a gear train to the camshaft and auxiliaries. The gear below the crankshaft gear drives the oil pump, while the gear on the opposite side of the engine from the camshaft drives the compressor and injection pump, which are mounted in tandem, a common arrangement on an engine of this type. Forward of the gear train is a multi-V-belt drive with a torsional vibration damper formed integrally with the crankshaft pulley.

Since this a DI engine, the combustion chamber is formed by a cavity in the piston, in this case a toroidal cavity. The injectors are inclined in the head, and the turbocharger is mounted above the exhaust manifold and halfway along the engine. There are very thorough cooling and lubrication systems, the coolant being directed at the sides of the liners, and then through holes into the cylinder head to play on to the injectors.

For lubrication, oil drawn from the sump is delivered by a gear pump to the system where first it passes through an oil-to-water cooler incorporated in the engine cooling system, then through the main filter to the main bearings and turbocharger. There is also a direct feed to jets that spray oil on the undersides of the piston crowns to assist in cooling. However, to ensure that the bearings always receive plenty of oil, there is a check valve which prevents the flow of oil to the jets until the pressure reaches a certain level.

The system also includes a by-pass filter. The oil pump is designed to deliver more oil than is needed, and some of the excess is fed through the by-pass filter instead of to the main lubrication system. Because this oil flows back to the sump, the

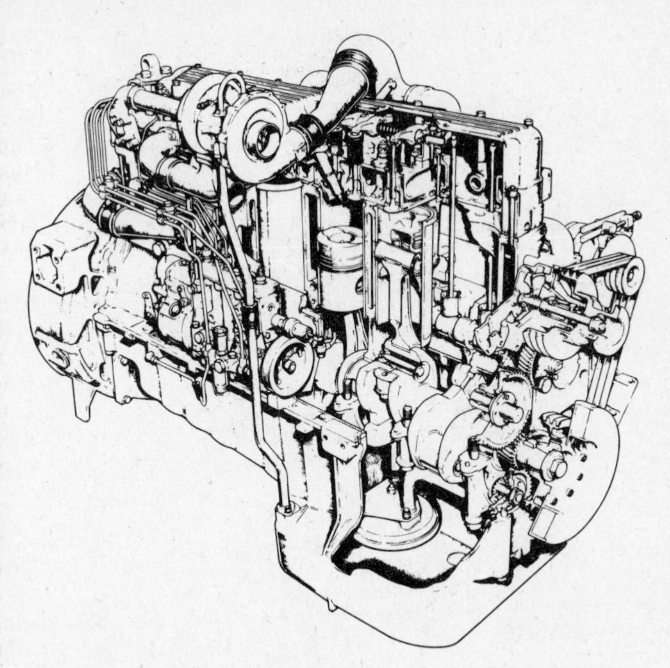

Fig. 17. Leyland TL12 turbocharged six-cylinder engine

by-pass filter can offer greater resistance than the main full-flow filter, and it can therefore trap smaller particles. This combination of full-flow and by-pass filters results in very good filtration.

. . . and the Fiat 8280 V8 diesel engine?

The Fiat 8280 is one of the largest diesel engines in use in commercial vehicles. With a bore and stroke of 145 × 130 mm this engine is unusual in that the bore is larger than the stroke. The displacement is 17.2 litres and maximum power output of 247 kW at 2400 rev/min. Maximum torque is 1130 N m (832 lbf ft) at 1200 rev/min.

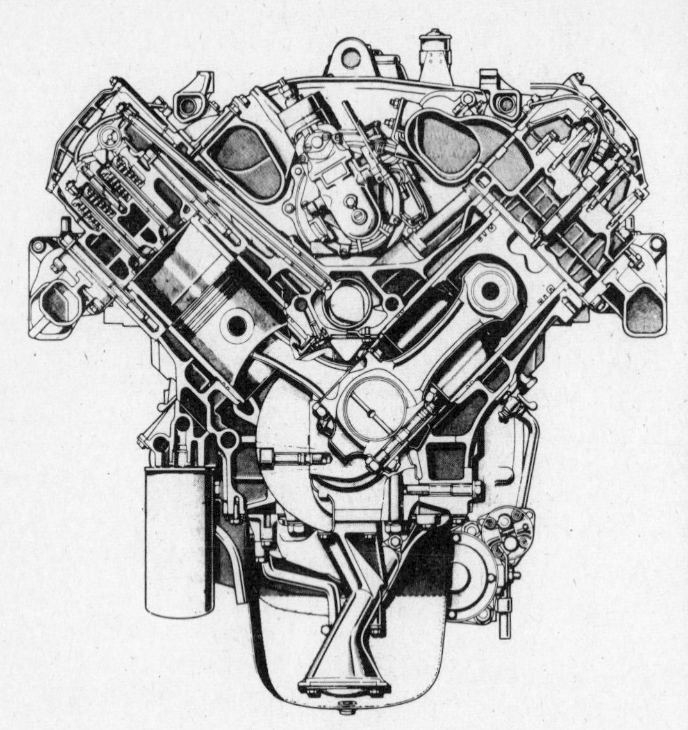

Fig. 18. The Fiat 8280 V8 engine of 17.2 litres is short but quite wide

Usually, the stroke of a diesel is larger than the bore, especially on large engines, in order to give more time for combustion. However, the use of a short stroke on a V-engine minimises the width. In this case, the large bores have allowed the use of four valves in the cylinder head, and a central injector. The injector should preferably be at the centre of the piston cavity, and this is easily accomplished with a central injector.

The construction of this engine is typical of a large heavy-duty diesel in that the cylinder block is a massive casting with wet cylinder liners. Each cylinder has an individual head retained by eight bolts.

Another interesting feature is that the helical gear timing train is at the rear of the engine. The camshaft actuates the valves through pushrods and rockers, one forked rocker actuating two valves. To keep the pistons cool, oil is sprayed to the undersides of the crowns.

. . . and the Cummins diesel engine?

The Cummins diesel engine features wet cylinder liners, a shallow, open combustion chamber and direct injection in which the injectors are actuated by separate valve gear. In addition,

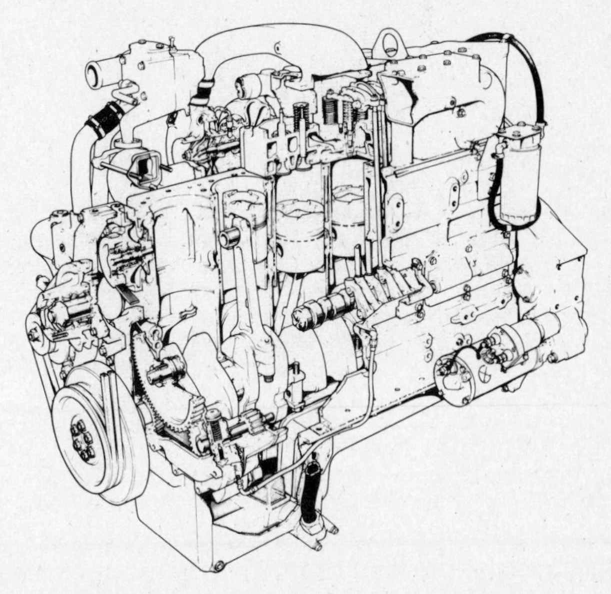

Fig. 19. The Cummins in-line engine with four valves per cylinder and central injector. Wet liners are used

there are four valves per cylinder, with a centrally-mounted injector.

The fuel is delivered by a low-pressure pump and governor to the injectors where the pressure is obtained by a pushrod and rocker mechanism. A number of holes in the injector nozzle direct the fuel at a shallow angle into the combustion chamber. The use of four valves per cylinder ensures high volumetric efficiency.

Cummins has concentrated on the manufacture of large engines in the 8–50 litres range for heavy-duty use. Among its automotive units, the compact V8s of 8.3 and 9.1 litres and the big in-line sixes of 14 litres are most commonly used. The V8s have very short strokes – a common feature on V-engines, but one which many designers consider far from ideal – with dimensions of 117 mm bore and 95 or 105 mm strokes (8.3 and 9.1 litre units respectively). Power outputs range from 147 kW at 3000 rev/min for the naturally-aspirated V-504 to 168 kW at 3000 rev/min for the 9.1 litre turbocharged VT-555 unit.

The 14 litre units, which are widely used in American and British trucks, are available in a number of different specifications. There are naturally-aspirated and turbocharged units with outputs of 179 kW. Since the turbocharged unit delivers maximum torque of 1085 N m at 1300 rev/min compared with 895 N m at 1500 rev/min, and maximum power is 200 rev/min lower than the naturally-aspirated version at 1900 rev/min, it is much more economical in use. Other turbocharged engines with power outputs of 216, 261 and 276 kW are available.

3

FUEL-INJECTION PUMPS

What are the basic types of fuel injection system?

(1) Jerk pump.
(2) Distributor pump.
(3) Low pressure system with unit injectors.

Do these types give different performances?

Yes, although to some extent they are interchangeable at the design stage. The jerk pump is the oldest design, is costly to manufacture, and may have slight variations in the pressures at different cylinders. However, it is capable of generating the very high pressures needed in larger diesel engines. It is not convenient for there to be automatic adjustment of timing in the jerk pump.

These comments apply generally to the unit injector system as well, except that in this case it is definitely impractical to alter the injection timing automatically.

The distributor pump is compact, relatively inexpensive to manufacture, and the pressure and amount of fuel delivered to each cylinder are identical. Timing adjustment can be accommodated easily. However, there is a limit to the amount of fuel that can be handled by this type of pump.

What is the jerk pump system?

In the jerk pump, a cam actuates a plunger to generate the pressure necessary to open the injector nozzle to allow fuel to be injected into the cylinder. There must be one cam/plunger assembly for each cylinder of the engine. To vary the amount of fuel delivered according to load and speed, a port in the barrel is open. This barrel can be rotated so that the port height is altered according to load and speed, to alter the amount of fuel delivered. However, the actual stroke of the plunger remains constant throughout.

What is the distributor-type fuel injection pump?

In this type of pump, exemplified by the CAV DPA and the Bosch VE, there is normally only one pumping element regardless of the number of cylinders in the engine. The charge of fuel is distributed through different discharge valves and pipes to the different cylinders by means of a rotary distributor. In some cases, there are two pumping elements, this configuration being adopted for powerful six- or eight-cylinder engines requiring a large amount of fuel to be delivered at rated speed and load.

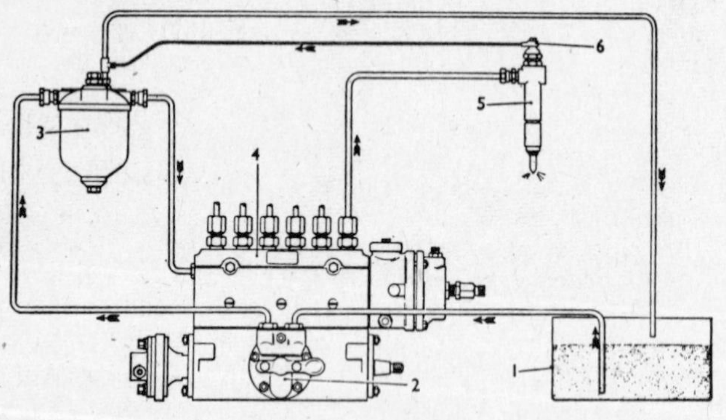

1. Fuel tank
2. Fuel-feed pump
3. Fuel filter
4. Fuel-injection jerk pump
5. Fuel injector
6. Injector leak-off union

(*Vauxhall Motors Ltd.*)

Fig. 20. Typical fuel-injection system layout

What does a typical fuel-injection system consist of?

A representative layout is shown in *Fig. 20*. This comprises a fuel tank, fuel-feed pump, fuel filter, injection (jerk) pump, fuel injectors and pipe lines including leak-off and excess-fuel return pipes.

How does the fuel-feed pump operate?

On many engines, the fuel-feed pump is mechanically-operated from an eccentric on either the engine camshaft or injection-pump camshaft. The diaphragm-type feed pump shown in *Fig. 21* is operated by an eccentric on the injection-pump camshaft.

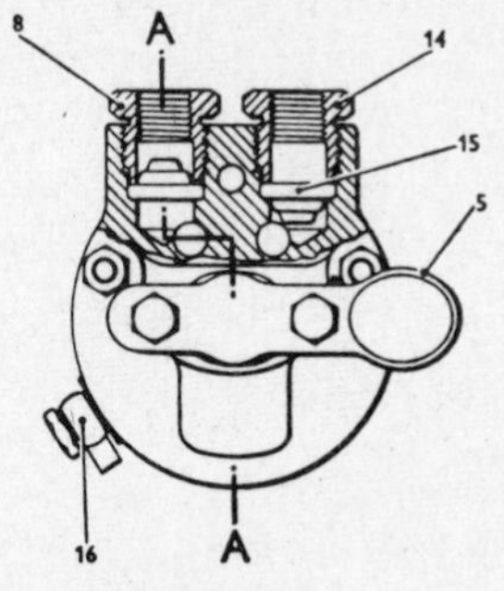

SECTION A–A

1. Feed-pump body
2. Feed-pump cover
3. Priming-lever spring
4. Spring cap
5. Priming lever
6. Priming spindle
7. Priming-spindle sealing ring
8. Fuel-outlet union
9. Outlet valve
10. Diaphragm
11. Diaphragm spring
12. Operating lever
13. Pivot pin
14. Fuel-inlet union
15. Inlet valve
16. Leak-off connection

Fig. 21. CAV diaphragm-type fuel-feed pump operated by an eccentric on the injection-pump camshaft

When the eccentric (shown dotted) is at its lowest point, it moves the pump operating lever (12) downwards, so drawing the diaphragm (10) inwards against spring pressure. The suction effect of the diaphragm draws the fuel oil into the pump from the fuel tank via the inlet valve (15). The return stroke of the diaphragm is dependent purely on the diaphragm spring, the strength of which determines the maximum delivery pressure, via the outlet valve (9), to the main fuel filter.

What is the function of the fuel-injection pump?

The function of the fuel-injection pump is to supply the engine with fuel in quantities exactly metered in proportion to the amount of power required and timed with the utmost accuracy, so that the engine will be smooth running and will deliver its output with the greatest economy.

As an example, the injection pump on high-speed engines must be capable of measuring accurately a volume of fuel oil of, say, less than one-fifth of a cubic centimetre and supplying this quantity of fuel, at a high pressure, to the fuel injectors many times per second. On one high-speed, two-stroke vehicle engine, at maximum speed, 40 injections of fuel are required from each pump element every second.

What is the purpose of leak-off or fuel-return pipes?

To return excess fuel from the fuel pump and injectors to the fuel tank. In the system shown in *Fig. 20*, excess fuel collected in the fuel-injection pump passes to the leak-off connection on the fuel-feed pump (see *Fig. 21*) through drain holes on the inner face and in the base of the feed-pump body.

How is the jerk-injection pump usually operated?

By means of a camshaft driven by the engine. In most cases, the camshaft is fitted in the base of the pump; in some engines, the pump operates from a camshaft forming part of the engine itself. The pump camshaft is driven at half engine speed for four-stroke engines and at engine speed for two-stroke engines. A separate pumping element is provided for each cylinder in the engine.

How are the pumping elements driven in the jerk pump?

Each pumping element contains a spring-controlled plunger. An extension from the end of this plunger rests on a cam so that

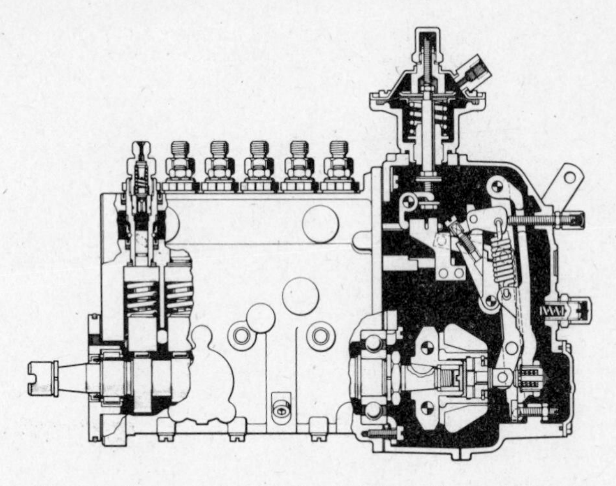

Fig. 22. Cut-away view of Bosch MW in-line fuel-injection pump

when the camshaft is driven from the engine, the plungers are given a reciprocating motion in correct sequence, corresponding to the firing order of the cylinders to which each pump element is connected.

What controls the amount of fuel oil injected into each cylinder?

The method which has been adopted in the majority of jerk-injection pumps is to keep the stroke of the plungers the same, but to provide means whereby a greater or less length of the stroke can be rendered ineffective. Thus on full load, the full stroke of the plunger would be utilised, whilst on half load the final half of the stroke would be ineffective because the fuel is allowed to escape through a spill port in the plunger casing. Thus though the stroke of the plungers is kept constant their *effective* stroke can be varied at will.

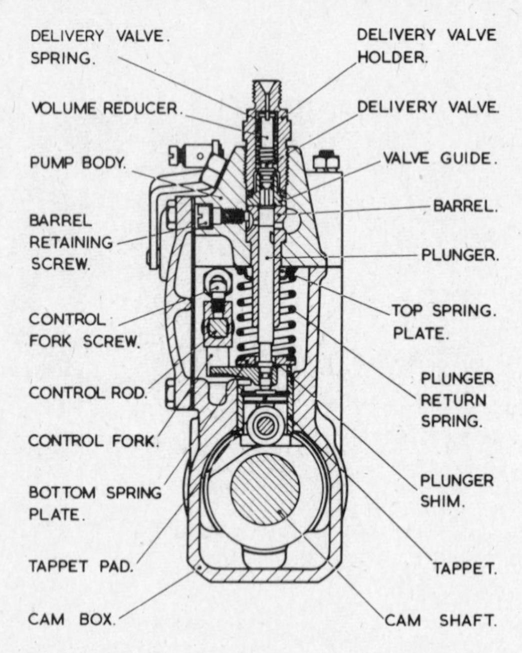

Fig. 23. Section through a typical jerk pump

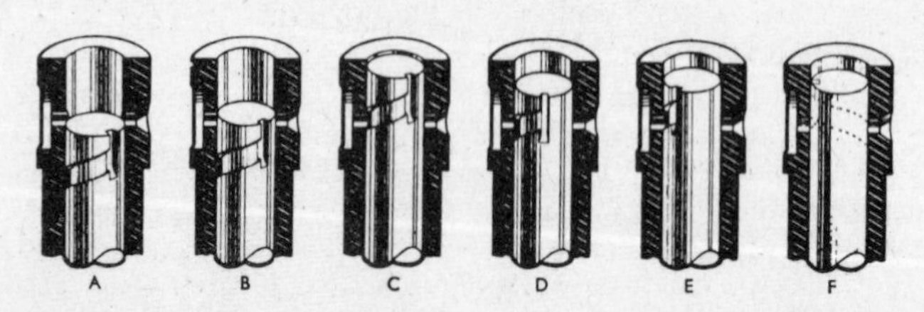

Fig. 24. Cut-away view of CAV jerk-injection pump barrel showing various plunger positions (see text for explanation)

How does a typical jerk-injection pump work?

The operation of a port-controlled, constant-stroke plunger is shown in *Fig. 24*.

When the plunger is at the bottom of its stroke (*Fig. 24(A)*) fuel under pressure in the pump gallery flows through the two ports to fill the interior of the barrel. As the plunger moves upwards, some of this fuel is forced out of the ports, until the plunger reaches the position shown at (B), when both ports are covered. At this point further upward movement of the plunger increases the pressure on the fuel and causes the delivery valve (mounted on top of the pump barrel) to be opened, and the fuel enters the pipe connected to the fuel injector.

The pipe and drillings in the injector are kept constantly filled, by previous operations of the plunger and delivery valve, and the extra fuel forced in raises the pressure in the pipe until, at the specified pressure, the injection valve is lifted off its seat. This enables the fuel to be discharged as an atomised spray from the hole or holes in the injector nozzle and penetrate the compressed air charge in the combustion chamber.

The fuel discharge continues until the helical edge of the plunger recess uncovers the spill port (see *Fig. 24(C)*), when fuel in the barrel flows down the vertical slot in the plunger and returns through the spill port to the fuel gallery.

The resulting drop of pressure in the barrel allows the delivery-valve spring to return the valve to its seat. In closing, the delivery valve draws a small quantity of fuel out of the pipe connected to the injector. This reduces the residual pressure in the pipe and enables the injector valve to snap quickly on its seat, thus preventing 'dribble' into the combustion chamber.

How is the effective stroke of the jerk-pump plunger varied?

The effective stroke of the plunger is varied by the movement of the pump control rod, which simultaneously rotates the plungers within the barrels, so that a wide or narrow section between the top of the plunger and the helical groove is in alignment with the barrel port. Commencement of fuel delivery is, therefore,

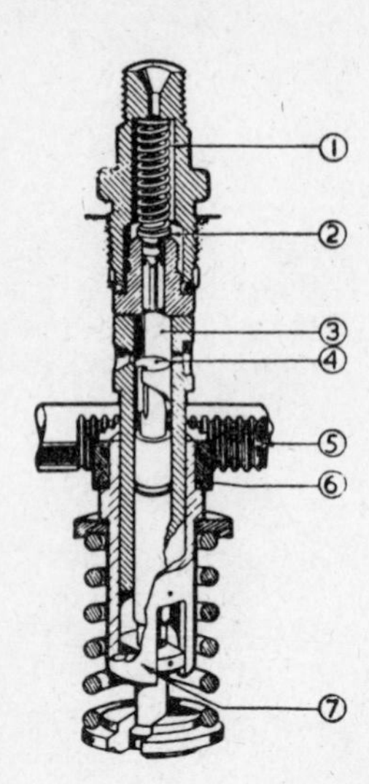

1. Valve spring
2. Delivery valve and seat
3. Pump barrel
4. Pump plunger
5. Control rod
6. Toothed quadrant
7. Control sleeve

Fig. 25. CAV jerk-injection-pump element in section

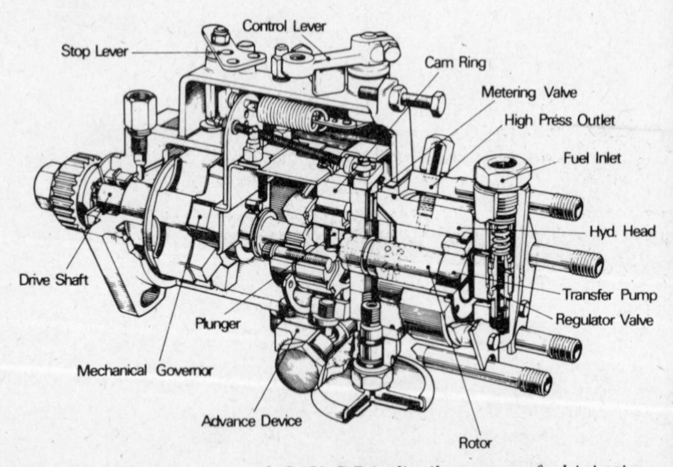

Fig. 26. Cut-away view of CAV DPA distributor-type fuel-injection pump with hydraulic governor

constant, but the end of the delivery stroke will depend on the load and speed at which the engine is operating. At (C), (D) and (E), *Fig. 24*, a plunger is shown in the position for full load, half load and idling speeds respectively, whilst at (F) the plunger is in the position required to stop the engine.

What is the main difference between the jerk pump and distributor pump?

In the fuel-injection jerk pump, a separate pumping element is used for each cylinder, whereas in the distributor pump the fuel is pumped by a single element and is distributed to each cylinder in turn by means of a rotary distributor.

What are the main features of the DPA distributor pump?

The main features of the CAV DPA fuel-injection pump are shown in *Fig. 26*. In this pump a central rotating steel member known as the pumping and distributor rotor is driven by splines from a drive shaft carried in the base of the pump housing, and it carries at its outer end a vane-type fuel transfer pump. The rotor is a close fit in a stationary steel cylindrical body, called the hydraulic head.

The pumping section of the rotor has a transverse bore containing two opposed pump plungers operated by a stationary internal cam ring carried in the pump housing, through rollers and shoes sliding in the rotor. Normally the cam ring has as many internal lobes as the engine has cylinders. The opposed plungers have no springs, but are moved outwards by fuel pressure.

The distributing part of the rotor contains a central axial passage which connects the pumping space between the plungers with ports drilled radially in the rotor which provide for fuel inlet and delivery. One radial hole is the distributing port and as the rotor turns this aligns successively with a number of outlet ports (equal to the number of engine cylinders) in the hydraulic head, from which the injectors are fed via external high-pressure pipes. A similar number of inlet ports spaced round the rotor align successively with a single port in the head. This is the inlet or metering port, and admits fuel under the control of the governor.

How is the fuel metered in the DPA distributor pump?

Fuel entering the pump through the fuel-oil inlet (*Fig. 26*) on the pump end plate, passes through a nylon filter to the inlet side of the vane-type transfer pump.

The fuel pressure is then raised to an intermediate level, known as 'transfer pressure', which is controlled by a piston-type regulating valve housed in the end plate. Transfer pressure does not remain constant but increases with the speed of rotation of the pump. Fuel at transfer pressure then passes through a passage in the hydraulic head to an annular groove in the rotor and thence to a chamber which houses the metering valve. The metering valve is operated by the engine-throttle control and regulates the flow of fuel through the metering port into the pumping section of the rotor.

The volume of fuel passing into the pumping element is thus controlled by the transfer pressure, the position of the metering valve and the time during which an inlet port in the rotor is aligned with the metering port in the hydraulic head.

How is the fuel pumped and distributed in the DPA distributor pump?

Pumping and distribution of the metered fuel are illustrated in *Fig. 27*. The left-hand diagram shows the charging phase and the right-hand diagram the actual pumping and distribution of the metered charge.

As the rotor turns, a charging port in the rotor is aligned with the inlet port in the hydraulic head and fuel at metered pressure flows into the central passage in the rotor and forces the plungers apart. The amount of plunger displacement is determined by the amount of fuel which can flow into the element while the ports are aligned. See left-hand diagram of *Fig. 27*.

The inlet port closes as rotation continues, and as the single distributor port in the rotor comes into alignment with one of the

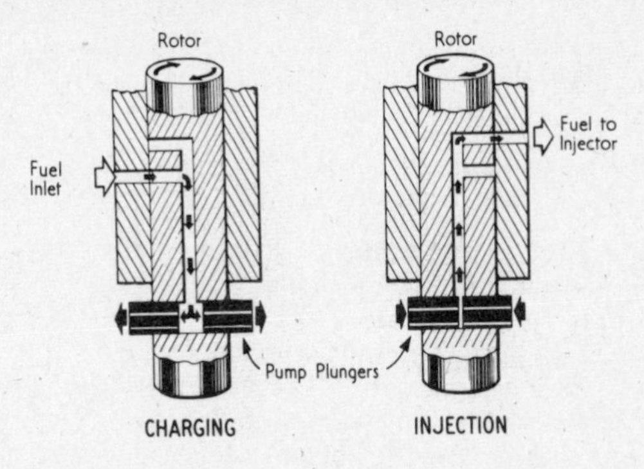

Fig. 27. Charging and injection cycles of the CAV DPA distributor-type injection pump

distributor ports in the hydraulic head, the actuating rollers contact the cam-ring lobes, and the plungers are forced inwards as shown in the right-hand diagram. High pressure is generated and fuel passes to the injector.

Are any special features incorporated in the DPA pump to suit use in cars?

Special versions of the DPA pump are produced for use in cars, and these are fitted to a number of European diesels. Additional bearings can be incorporated to suit the toothed-belt drive, while extra fuel can be delivered to help with cold starting. This excess fuel device takes the form of an extra pair of pumping plungers operating only at cranking and low speeds. These elements are disengaged as the speed rises. Also incorporated are a device to advance the injection timing at idling, and a combined speed and load control for the automatic advance system. Other features available include a two-speed governor which controls idling and maximum speed only, although standard mechanical or hydraulic governors are also available. A solenoid stop valve between the transfer pump and metering valve is also incorporated to allow the 'ignition key' to be used to stop and start the engine.

What is the unit injector system?

Although unit injectors are used on a number of American engines, including the Detroit Diesels, the American Cummins system is probably the best known, and it is a good example of this system. In the Cummins PT (pressure time) system (*Fig. 28*) there is a fixed opening in the injector, the pressure and time to meter the fuel charge dictating the amount of fuel injected. The basic elements in the system are the fuel pump, governor,

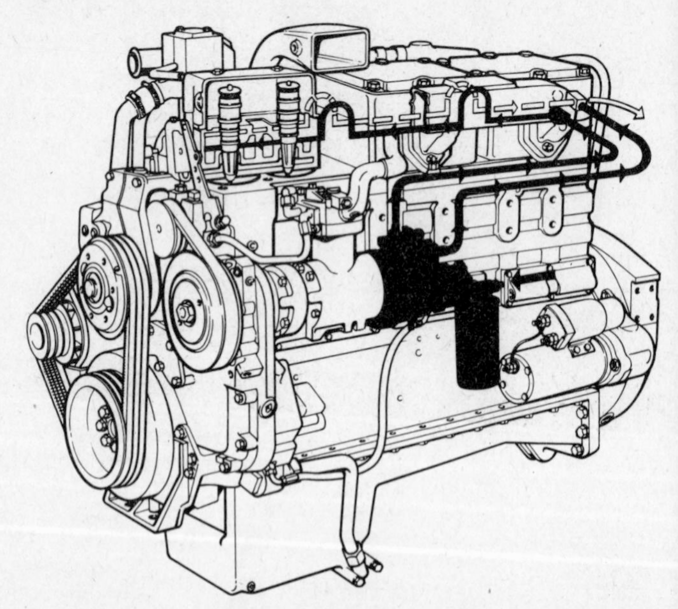

Fig. 28. The Cummins PT fuel-injection system is based on a low-pressure pump and governor unit and a high-pressure injector

throttle and camshaft-actuated injector. The fuel pressure is regulated by the governor/fuel pump, while the engine speed determines the length of opening of the injector – as speed increases, so the duration of opening decreases.

How does the fuel circuit in the Cummins PT system operate?

Fuel is drawn from the tank through a fine filter to the gear-type fuel pump which is driven by the engine through a train of gears. The pump does not have to be timed relative to the engine. Built into the pump is a governor/throttle, and a magnetic screen filter protects the system. Then, a single fuel manifold takes the fuel to the injectors, while a return line takes it from the injectors back to the tank. Each injector is actuated by a special cam (on the engine camshaft), pushrod and rocker to pressurise the fuel and deliver it to the cylinder.

There is an electric solenoid valve to stop the fuel being delivered to the injectors when the key is switched off.

How does the fuel pump/governor operate?

The pump is a low-pressure unit, but of course, pressure increases with speed to give basic metering. There is a flyweight-type governor to control both idling and maximum speeds. Then, the fuel passes diametrically through a throttle shaft which is connected to the accelerator pedal. Rotation of the throttle shaft alters the amount of flow that can pass to the injectors and the fuel pressure. At idling the main fuel supply is cut off by the throttle shaft, some fuel being delivered through idling passages to the engine.

How does the injector operate?

Basically the injector consists of a body inserted into the cylinder head, and a spring-loaded plunger whose conical nose sits in a

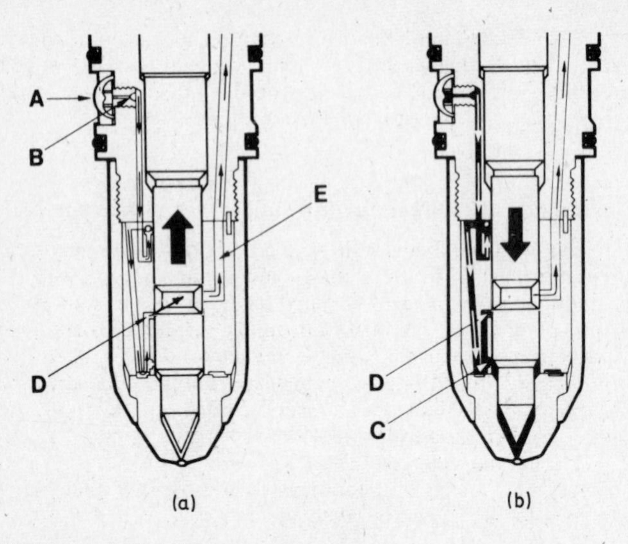

(a) *Start upstroke (fuel circulates):*
Low pressure fuel enters injector at (A), flows through inlet orifice (B), internal drillings, around annular groove in injector cup, and up passage (D) to return to tank. Engine speed, governor and throttle determine fuel pressure which, before inlet orifice (B), determines amount of fuel flowing through injector.
(b) *Upstroke complete (fuel enters injector cup):*
Injector plunger moving upwards uncovers metering orifice (C) and fuel enters injector cup. The amount is determined by fuel pressure. Passage (D) is blocked, momentarily stopping fuel circulation and isolating metering orifice from pressure pulsations.

Fig. 29. The Cummins PTD fuel injector in operation

seat at the injector cup. Raising the injector plunger, following the previous injection stroke, allows fuel at low pressure to be delivered through passages in the body, around an annular groove in the plunger shank, and back to the return passage (*Fig. 29*). Completion of the upstroke uncovers the metering orifice C, and fuel enters the injector cup, the amount depending on the fuel pressure, while the outlet passage D is temporarily blocked.

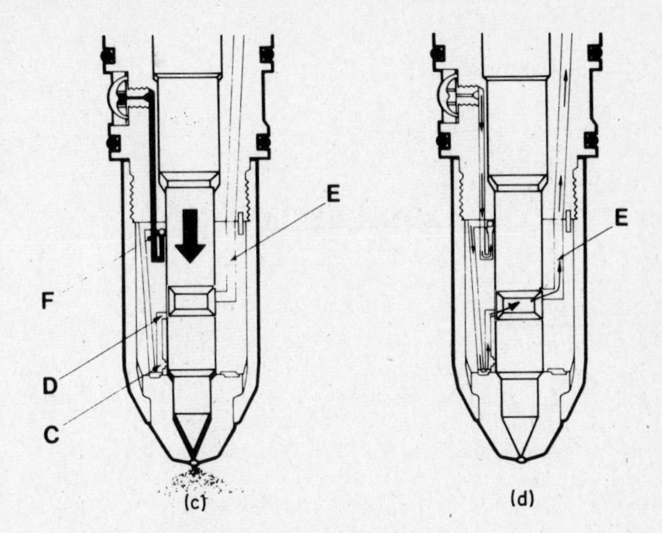

(c) *Downstroke (fuel injection):*
Plunger moving downwards closes metering orifice and cuts off fuel supply to cup. Continuing down, plunger forces fuel at high pressure from cup through tiny holes as fine spray. This assures complete combustion of fuel in cylinder. When plunger undercut uncovers passage (D) fuel again flows through return passage (E) to tank.
(d) *Downstroke complete (fuel circulates):*
After injection, plunger remains seated until next metering and injection cycle. Although no fuel is reaching injector cup it flows freely through passage (E). This cools the injector and warms fuel in the tank.

As the plunger moves down, so the fuel supply is cut off, and the fuel in the spray cup is pressurised and delivered as a fine spray into the engine cylinder. After injection, the plunger remains seated in the injector cup, but the fuel can flow freely through the injector and back to the fuel tank.

4

GOVERNORS

How is the speed of diesel engines controlled?

Some form of governor is always fitted to control the speed of the engine. For constant-speed applications, the governor controls the speed automatically within specified limits under changes of load between no load and full load.

In automobile engines, speed control is by a combination of governor and accelerator pedal operated by the driver of the vehicle. Locomotive and marine engines employ governor control combined with manual control.

Why is a governor required when under driver control?

Unlike the petrol engine, the diesel engine is not self-regulating in respect of either load or speed. That is to say, as the load on the engine is increased the amount of fuel may be continually increased beyond that which can be burned in the oxygen available, resulting in excessive carbon formation, heavy exhaust smoke and high exhaust temperature, due to late burning of the charge. This can only lead to permanent damage to the engine components and must be avoided, the usual method being by the use of a maximum-fuel stop which is set at the normal safe engine-load position.

In the same way, if the engine load is reduced or removed, engine speed will increase unchecked, and as the only limit to the amount of fuel injected is the maximum-fuel stop just mentioned, very dangerous engine speeds may be reached. This state of affairs is aggravated by the fact that most injection pumps have what is known as a 'rising characteristic'; that is, at a fixed rack setting the pump output increases with increasing speed. This is due to 'wire-drawing' at the sleeve ports and the net result is that the faster an unloaded engine runs the faster it wants to go.

On the other hand, at idling conditions, as the engine loses speed the pump output decreases and the engine will stop.

To meet these conditions, a diesel engine is fitted with a governor which take one of several forms.

What in general terms is the operation of the diesel engine governor fitted to jerk pumps?

The governor control rod is connected to the fuel-pump rack. When the engine speed tends to rise, the governor mechanism operates on the injection-pump rack to turn the pump plungers slightly, thus shortening the effective stroke of the pumps.

This has the effect of reducing the amount of fuel injected into the engine cylinders so that the tendency of the engine to speed up is checked.

If the speed falls below normal, the fuel-pump rack is moved by the governor mechanism so as to increase slightly the fuel supply to the engine cylinders.

What forms of governor are in use?

Mechanical (i.e. centrifugal), pneumatic and hydraulic.

What is the action of a simple constant-speed mechanical governor?

Referring to *Fig. 30*, it will be seen that the governor spindle is driven from the engine shaft through suitable gearing. On the upper end of the governor spindle is mounted a pair of governor weights connected by a pair of springs which pull the weights towards the spindle as the governor slows and comes to rest.

Each governor weight is located at one end of a bell-crank lever, the other ends of the bell-crank levers bear on a collar which forms part of a sleeve mounted on the governor spindle.

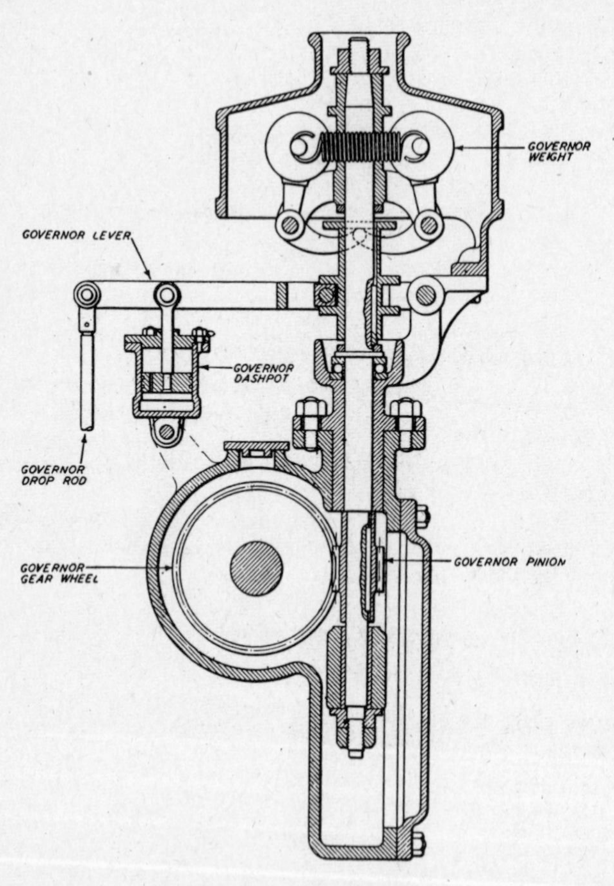

Fig. 30. Simple constant-speed mechanical governor. The drive is taken from the engine through suitable gearing (Petters Ltd.)

When the governor spindle begins to revolve, these weights tend to move outwards. This outward motion of the balls causes the lower ends of the bell-crank levers to exert an upward thrust on the movable sleeve which is mounted on the governor spindle.

As the sleeve moves upwards it causes the governor lever to pivot on its fulcrum, which is located in a bracket on the right of the governor casing. The movement of the governor level is transmitted through suitable link-work to the control rack of the fuel pump.

If the engine speed increases, the governor balls fly farther apart and so raise the sleeve and through this the governor lever, which through its associated link-work reduces the fuel supply to the engine. If the engine speed drops, the reverse action takes place. It will be seen that in this particular design of governor a dashpot is used to prevent 'hunting'.

What types of mechanical governor are fitted on automobile engines?

The idling and maximum-speed governor, the all-speed governor fitted to jerk-injection pumps, and the all-speed governor fitted to the DPA distributor-type injection pump.

What is the action of the idling and maximum-speed mechanical governor?

The idling and maximum-speed governor prevents the engine from either stalling or exceeding a predetermined maximum speed, but otherwise has no part whatsoever in speed regulation. Any speed and load of the engine between the idling and maximum-speed settings of the governor is controlled directly by the operator by means of the accelerator pedal.

What is the basic operation of a mechanical governor?

Mechanical governors are generally used on automotive diesels, and these incorporate flyweights. A simplified governor system, a CAV unit of particularly modern design, is shown in *Fig. 31*.

The control rod (1), for the rack that actually alters the amount of fuel delivered, is connected through a lever (2) to a sleeve (3) on the governor shaft. The governor shaft itself is formed at the end of the camshaft (4) and carries a cage containing flyweights (5). These react on a shoulder on the sleeve, another shoulder (6) on that sleeve (3) being acted on by leaf springs (7). The accelerator pedal is connected to a bell-crank lever (8). At the free end of this bell-crank lever is a fork and roller which bears against a ramp (9).

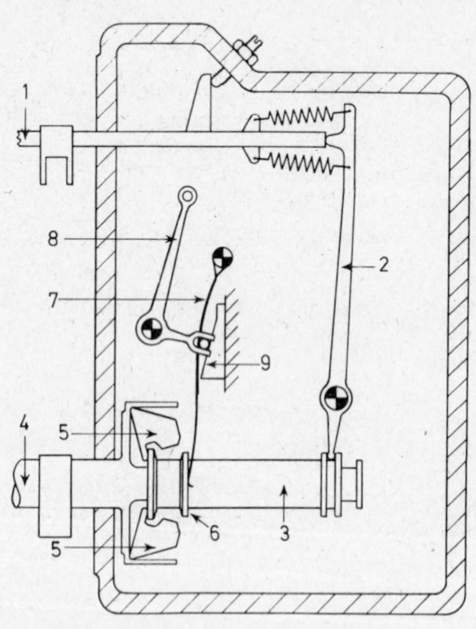

1. Rack control rod
2. Lever
3. Sleeve
4. Camshaft
5. Flyweights
6. Shoulder
7. Leaf spring
8. Bell-crank lever
9. Ramp

Fig. 31. The CAV C governor in simplified form

As the engine speed rises, so the flyweights (5) pivot in the corners in the cage and push the sleeve along the shaft. Therefore, the lever (2) actuates the control rod to reduce the amount of fuel delivered. However, depression of the accelerator pedal actuates the bell-crank lever so that the roller runs up the ramp and increases the force applied to the leaf springs. This force opposes that of the flyweights on the shoulder. Therefore, the pump fuelling is a function of the forces exerted by the weights and the governor springs.

How does this control maximum speed and idling?

At maximum speed, the force exerted by the flyweights on the sleeve exceeds that exerted by the spring, so the rod is moved to reduce the amount of fuel supplied to the engine. At idling, the forces applied by the spring and the flyweights work against each other to keep the engine running.

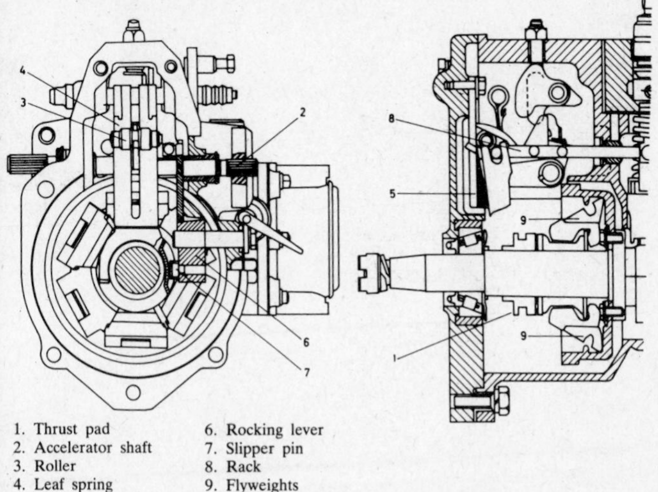

1. Thrust pad
2. Accelerator shaft
3. Roller
4. Leaf spring
5. Ramp
6. Rocking lever
7. Slipper pin
8. Rack
9. Flyweights

Fig. 32. The CAV GCV governor in detail

What is the construction of such a governor?

In practice, the construction is more complex than in *Fig. 31*, a typical unit being shown in *Fig. 32*. This is the CAV GCV governor in which the governor cage is carried on the extension to the camshaft, while the sleeve and thrust pad (1) are free to slide on the shaft extension. The complete governor unit is enclosed in an extension housing.

The accelerator shaft (2) is linked to the roller (3) by the bell-crank lever, and this is held between the leaf spring (4) and the ramp (5). A rocking lever (6) with slipper pin (7) connects from the thrust pad to the control rod that actuates the plunger barrels. The rack (8) actually alters the amount of fuel delivered.

It can be seen that to keep the governor compact the components are placed close together. The flyweights (9) seat in a detachable cage.

How can the characteristics of this type of governor be altered?

The characteristics can be altered in several ways, (i) by changing the number of weights used, (ii) by altering the travel and rocker ratio of the weights, (iii) by changing the gauge of the leaf springs and the slope of the ramp.

What is the layout of the all-speed mechanical governor fitted to the DPA distributor-type fuel-injection pump?

Where a mechanical governor is fitted to the DPA distributor-type injection pump, the pump housing is lengthened to accommodate the governor unit which is carried on the drive shaft. The shaft is carried in a substantial hub which has a journal bearing in the pump housing and carries the flyweight mechanism.

The governor flyweights are held in a carrier clamped between the driving hub and a step on the driving shaft. The weights are so shaped that in operation they pivot about one edge; as they move in or out under varying centrifugal force according to the pump speed, they operate a thrust-sleeve which slides on the

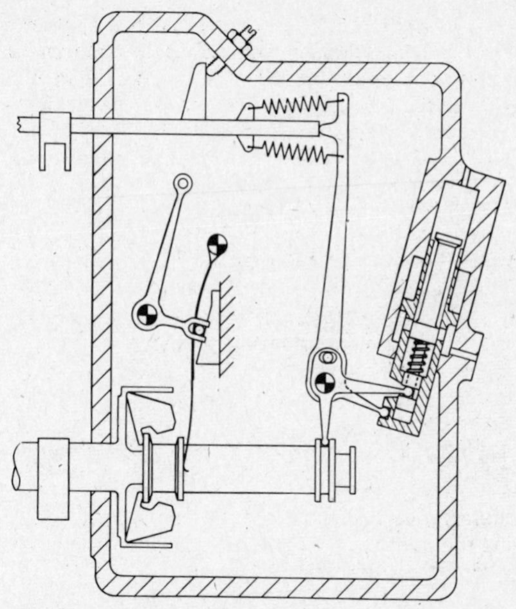

Fig. 33. Simplified construction of CAV CS governor with hydraulic servo

drive shaft, moving it axially to control the admission of fuel, through the governor mechanism, by means of the rotary metering valve.

A spring link connecting the governor lever to the metering valve enables the valve to be closed by the stopping mechanism without having to overcome the governor spring. Stopping may be by hand or by a solenoid-operated device.

How does the all-speed mechanical governor fitted to the DPA distributor-type pump operate?

The operation of this type of governor may be followed in detail be reference to *Fig. 34*. Flyweights (B) and thrust sleeve (A) are carried on shaft (T), the face of sleeve (A) being in contact with control arm (C). This arm is connected to the metering valve (O) by a spring-hook assembly (N); the spring is a light one and permits overriding movement of the valve to the shut-off position when stopping the engine.

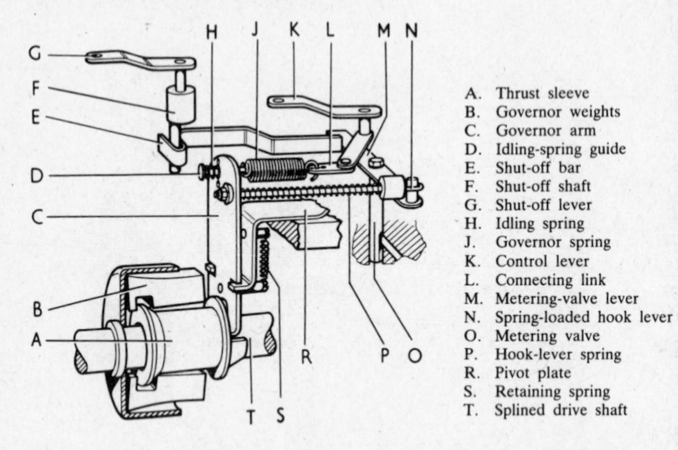

Fig. 34. Mechanical governor linkage of the CAV DPA distributor-type injection pump

Control arm (C) pivots about a knife-edge on the flat steel bracket (R). A flat steel fuel shut-off bar (E) is operated by a small cranked shaft (F), and links up with the metering valve. By operating the fuel shut-off lever (G), the shaft (F) will move bar (E), and no matter in what position the govenor weights and sleeve may be, this bar can turn the metering valve (O) against the pressure of the light spring (P), thus stopping the engine.

The governor is controlled by a spring (J), attached through linkage (L) which is connected to shaft and control lever (K). The spring is attached to the control arm through the idling-spring guide (D), which is spring-loaded by the light-idling spring (H). Engine control by lever (K) is therefore effected by adjusting the load on the governor spring (J), which opposes the thrust of the flyweights through arm (C). This is spring-loaded during idling by the light-idling spring, but as the speed increases, the main-control spring comes into action.

When starting, the throttle lever (K) is put hard over, holding the metering valve in the full-fuel position. As the engine fires the lever can be brought back and the governor then operates in the idling position.

Movement of the throttle lever adjusts the load on the governor-control spring, bringing about a change in the position of the control arm, and hence of the metering valve, admitting more or less fuel to the pump as required. For any setting, the governor will maintain the speed within close limits.

Tensioning of the governor spring provides increased resistance to the movement of the governor control arm under the influence of the governor weights, so that with greater tension, resulting from increased throttle opening, governor control will be at higher rev/min.

Within idling-speed range, tension is removed from the governor spring (J), and the light-idling spring (H) gives sensitive control at low rev/min.

What is the principle of the pneumatic governor?

It uses the depression in the engine induction manifold to control the fuel pump and thus the engine. Engine power is controlled at any given speed throughout the entire speed range between maximum and idling limits predetermined by stop-screw adjustments.

What is the construction of a pneumatic governor?

The pneumatic governor shown in *Fig. 35* consists of two main parts: (1) the venturi unit, located between the induction-manifold pipe on the engine and air filter; (2) the diaphragm unit, mounted on the fuel pump and connected to the fuel-control rack.

It will be seen that the body interior is shaped to form a venturi (B), the actual dimensions of which are determined by

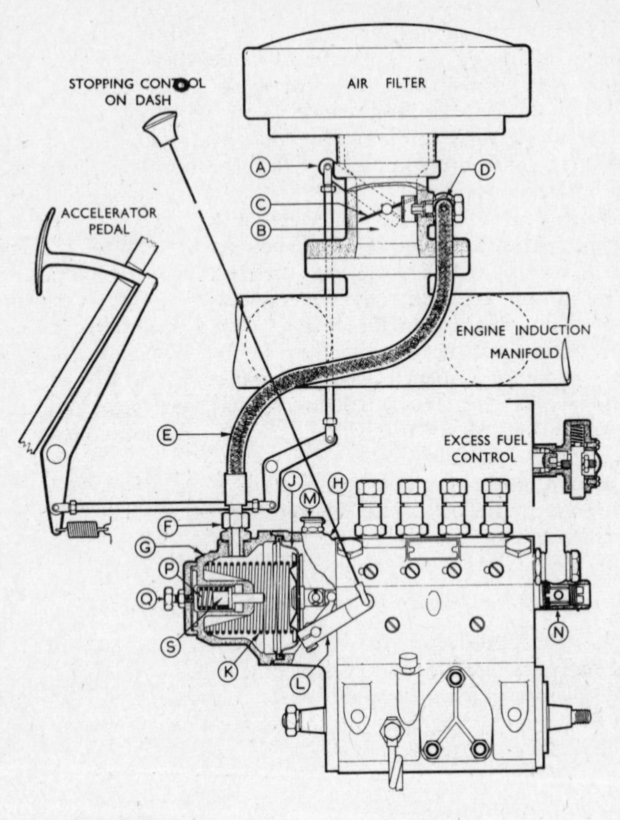

A. Venturi valve-control lever
B. Venturi throat
C. Venturi butterfly valve
D. Vacuum-pipe union
E. Vacuum pipe
F. Diaphragm-housing union
G. Diaphragm housing
H. Main housing
J. Diaphragm
K. Main diaphragm spring
L. Stop lever
M. Oil cap
N. Control-stop rod
P. Auxiliary idling spring
Q. Auxiliary idling set screw
S. Auxiliary idling plunger

Fig. 35. Pneumatic governor with single-pitot venturi in flow-control unit

the engine to which it is fitted and the speed range desired. A butterfly valve (C) is provided to control the air flow, and is connected through a spindle and lever (A) to the accelerator pedal. The movement of the butterfly valve for maximum and idling speeds can be controlled by adjusting stops (T) and (U) (see *Fig. 36*). A small auxiliary venturi situated within the main venturi may be secured to, or cast integrally with, the body of the venturi unit. Projecting into the auxiliary venturi at right angles to the airflow is a pitot tube which is connected to the diaphragm unit by the flexible pipe (E). (Double-pitot venturi flow-control units incorporate a second auxiliary venturi and pitot tube. This pitot tube is connected by a flexible pipe to an air valve situated in the diaphragm unit (see *Fig. 37*).)

The diaphragm unit consists of a main housing (H) and diaphragm housing, between which a diaphragm (J) is held, and thus an airtight compartment is formed within housing (G). Spring (K) is arranged in this compartment to act on the diaphragm, and tends to push the pump-control rod to the maximum-fuel stop (N), and lever (L) is provided so that the engine may be stopped by operation of a control attached to the lever by a wire or linkage and fitted on the dashboard.

How does the pneumatic governor operate?

Referring to *Fig. 35*, the operation of the governor is as follows. Spring (K) pushes the control rod to the maximum-fuel position, thus allowing sufficient fuel for starting the engine, but when the engine is running and the accelerator is returned to the idling position, the butterfly valve (C) remains practically closed, and a high vacuum is created in the vacuum pipe (E) and behind the diaphragm. Consequently, the pressure on the other side of the diaphragm, which is atmospheric pressure, overcomes the tension of the main spring (K), and the diaphragm draws the control rod towards the 'stop' position until the engine is running at the predetermined idling speed for which the idling stop (U), *Fig. 36*, has been set.

When the accelerator is depressed the butterfly valve is opened, and this causes a reduction in velocity of the air passing through the venturi throat, so that the vacuum is relieved, which allows spring (K), *Fig. 35*, to push the control rod towards the full-load position, to increase the amount of fuel pumped and raise the engine speed.

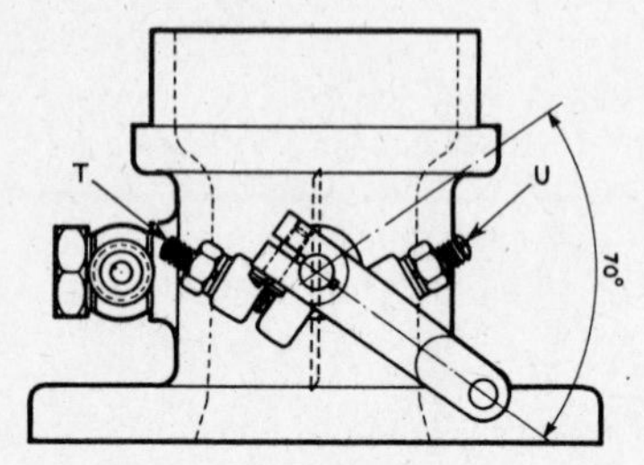

T. Adjustable screw for maximum speed
U. Adjustable screw for idling

Fig. 36. Control stops on venturi of CAV flow-control unit

When the operating-lever is against the maximum stop-lever (T), *Fig. 36*, the maximum engine speed is obtained. Should the engine speed tend to exceed the predetermined figure, an increased depression is created in the vacuum pipe and spring compartment, due to the increased velocity of the air through the venturi throat, which draws the control rod towards the 'stop' position and thus restores the engine speed to normal.

What methods of diaphragm control for idling speed are available?

Two methods are available when a single pitot-venturi unit is fitted, namely setscrew adjustment and cam adjustment.

With the setscrew adjustment (*Fig. 35*) it is possible to set and permanently fix the point at which an auxiliary or damping spring comes into operation, whereas with the cam type the auxiliary spring is tensioned by the cam (V), *Fig. 38*, which is connected to the accelerator or speed control through a spindle and lever (W). Only when the accelerator is returned to the idling position is the spring (P) brought into operation.

A. Butterfly valve
B. Auxiliary venturi
C. Diaphragm chamber
D. Secondary auxiliary venturi
E. Air valve
F. Diaphragm
G. Governor spring

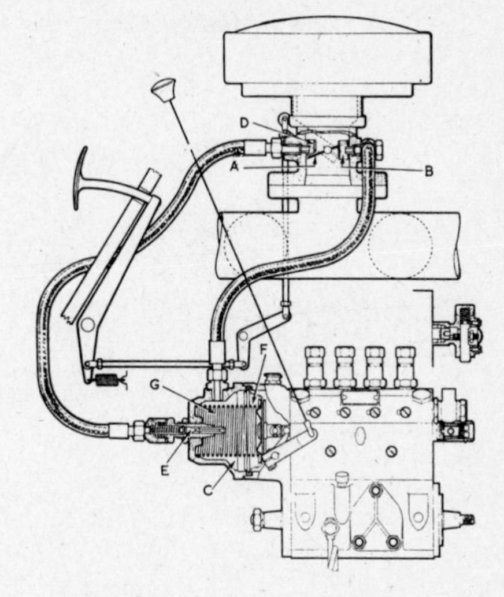

Fig. 37. CAV pneumatic governor with double-pitot venturi flow-control unit

P. Auxiliary idling spring
S. Plunger
V. Cam
W. Cam-operating lever

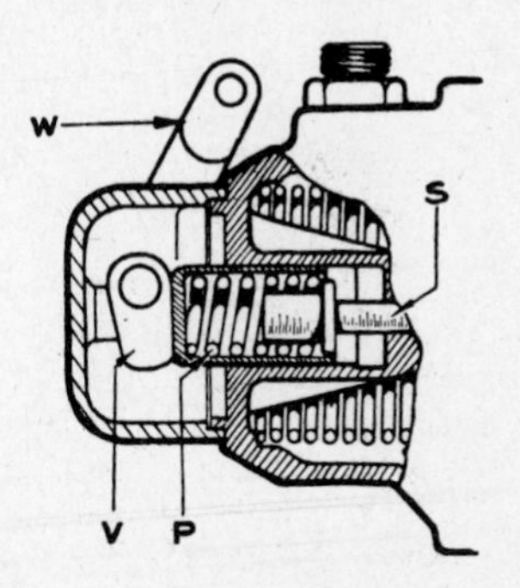

Fig. 38. Cam-operated auxiliary stop on CAV pneumatic governor

When a double-pitot venturi unit is fitted, the auxiliary-idling stop is replaced by a spring-loaded air valve (E), *Fig. 37.* As the diaphragm moves beyond the normal idling position, this valve is opened to permit relief of the depression in the diaphragm chamber.

How is the idling speed adjusted on the pneumatic governor?

When a single-pitot venturi is fitted, the idling speed is adjusted by movement of the idling stop (U), *Fig. 36.* The adjustment of the auxiliary-idling stop (setscrew or cam adjustment, as described above) should only be made when hunting occurs at the correct idling speed. If hunting does occur, ensure that it is not attributable to engine faults (such as faulty fuel-injectors) before making adjustments.

When a double-pitot venturi is fitted, the positions of the idling stop on the venturi and of the air-valve assembly in the diaphragm unit must be adjusted at the same time. Adjustment of either will affect idling speed, and it is only by adjusting both settings simultaneously that a setting can be achieved where even running is obtained at correct idling speed.

What is the function of an excess-fuel device?

This device permits overriding of the maximum fuel stop to provide excess fuel for starting.

How does a typical excess-fuel device operate?

Fig. 35 shows one type of excess-fuel device which is incorporated with the maximum-fuel stop. The illustration shows the maximum-fuel stop (in the maximum-fuel position) carried on a spring-loaded plunger. When the plunger is depressed, the stop screw is moved into alignment with a hole drilled in the end of the control rod, thus permitting the rod to move beyond the normal maximum fuel position. When the engine has started and the accelerator pedal has been released, the control rod moves towards the idling stop, and the plunger and the maximum-fuel stop are returned automatically to the normal working position by the spring.

What is the construction of the hydraulic governor fitted to 'in line' (jerk) injection pumps used on transport vehicle engines?

A representative type of hydraulic governor for controlling variable-speed oil engines of up to approximately two litres capacity per cylinder is the CAV H-type. It is of the all-speed type, enabling the engine speed to be set and maintained at any desired value between idling and maximum.

This governor operates on the principle of the 'inverted hydraulic amplifier', by which a small change in pressure in one part of the system produces in another part an opposite change of much greater magnitude.

The governor is mounted at either end of the injection pump. A maximum fuel stop, which also incorporates an excess fuel device for ease of starting, is carried in a separate housing mounted on the opposite end of the injection pump to that of the governor.

Safety is ensured by the employment of a shut-down lever that can be operated manually against the operation of the governor. In addition, the servo spring will take the control rod back to zero if for any reason the oil pressure fails.

The governor uses the normal fuel oil, drawn from the injection pump, as its operating medium. A duct connects the fuel-oil gallery of the pump with the inlet to the governor gear pump which provides the operating pressure.

The gear pump, mounted on the lower extension of the main casing, is of very simple construction employing two hardened and ground gears of normal involute form. It is driven from a sleeve on the camshaft through a self-aligning coupling to relieve the gears of side loading due to the drive, and the body is constructed in three parts. One part carries the bearings for the spindles, the intermediate portion (of cast iron and face-ground to close limits) contains the gears, and the assembly is completed by a plain aluminium cover, the three parts being dowelled and held together by studs passing right through to the mounting face. No gaskets are used at the joint faces.

All the moving parts of the governor are carried in inserted sleeves or bushes, those for the main pistons and plungers being fully floating to avoid distortion, and compressible synthetic rubber rings are employed to seal against leakage.

What is the operating principle of the CAV H-type hydraulic governor?

The gear pump (A), *Fig. 39*, picks up fuel oil from the fuel gallery of the injection pump and delivers it under pressure through a diffuser (D) to the amplifier chamber, from which the fuel oil escapes through the orifice in the amplifier piston (F). The pressure drop through the orifice will set up an endwise thrust on the amplifier piston, depending upon the amount of fuel oil flowing, i.e., upon the gear pump speed, which in turn is proportional to engine speed, since the gear pump is driven directly from the camshaft of the injection pump.

After passing through the amplifier piston, the fuel oil is led to the servo piston (G), which it moves against the servo-piston spring (H). This servo piston is coupled to the control rod of the injection pump, and the movement referred to has the effect of moving the injection-pump control rod towards the 'open' position and so increasing the pump delivery. The pressure generated at this point is limited by the high-pressure relief valve

---

Key to Fig. 39

A. Gear pump
A1. Vent plugs
B. Low-pressure relief valve
C. High-pressure relief valve
D. Diffuser
E. Amplifier valve
F. Amplifier piston
G. Servo piston
H. Servo-piston spring
J. Injection-pump control rod
K. Drag link
L. Swing link
M. Inner spring
N. Idling-valve inner plunger
O. Outer-spring
P. Idling-valve outer plunger
Q. Idling-valve adjusting screw
R. Inner control plunger
S. Outer control plunger
T. Control spring
U. Control pawl
V. Governor lever
W. Control shaft
X. Maximum-speed stop screw
Y. Control lever
Z. Idling-speed stop screw

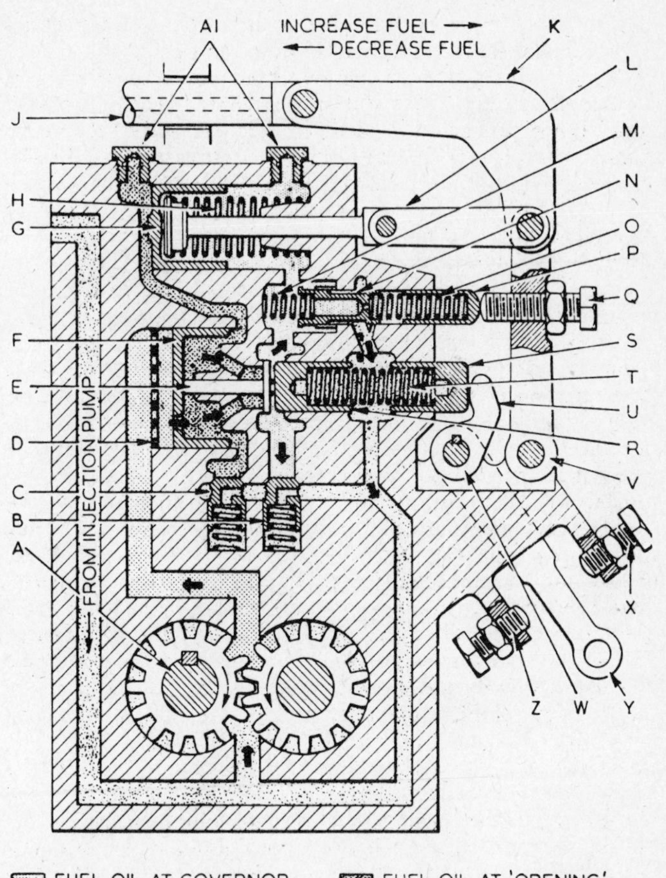

Fig. 39. Principle of operation of CAV H-type hydraulic governor (see facing page for key to letters)

(C). Excess of pressure causes fuel oil to be released through this valve to the inlet connection of the gear pump.

The end thrust on the amplifier piston, previously referred to, causes this piston to bear against the stem of the amplifier valve (E), tending to open this valve. This tendency is resisted by the load on the control spring (T). The control spring loading is varied by the control pawl (U), which is keyed to the control shaft (W), and operated by the accelerator-pedal linkage through lever (Y). Depressing the accelerator forces the outer plunger (S) towards the inner-control plunger (R), thus compressing the control spring (T) to increase the load on the amplifier valve. The amplifier valve will thus open at a pressure dependent upon the control pawl position selected by the driver. The fuel oil which flows through this valve is led to the inside of the servo piston, where it acts to assist the servo-piston spring (H) in opposing the 'opening pressure' and brings the servo piston to a state of balance, depending on the difference between the 'opening pressure' and 'closing pressure'.

This 'closing pressure' is limited by the low-pressure valve (B) which opens under excess pressure and returns fuel oil to the inlet side of the gear pump. Also in communication with the 'closing valve' is the idling valve (N). This valve can allow fuel oil to escape to the gear-pump inlet through slots in the idling-valve body which are opened or closed by a collar on the valve piston. As already mentioned, the idling valve is designed to give greater sensitivity under idling conditions than can be given by the amplifier alone, with the lower rates of fuel-oil flow obtained at idling rev/min.

It will assist the explanations of the operation of the idling valve if we first study the governor lever (V). This is pivoted at its lower end, and the upper end is coupled through a swing link (L) to the servo piston. The upper end also carries a drag link (K) which is connected to the injection-pump control rod. At an intermediate point the lever carries an adjusting screw (Q) against which the idling-valve outer plunger (P) bears. Thus the plunger (P) also has a motion proportional to those of the servo piston and control rod.

The two plungers (N) and (P) of the idling-valve assembly are held apart by the outer spring (O), while the spring (M) is placed between the inner plunger (N) and the governor housing, holding both plungers towards the governor lever and the adjusting screw.

A small orifice through the inner plunger (N) allows fuel oil at 'closing pressure' to fill the space between the plungers, but restricts the passage of fuel oil in or out of this space. In other words, a dashpot action is obtained.

For rapid movement of the control rod, governor lever and outer plunger (P), such as are usual under 'idling' conditions, the restriction causes the fuel oil to be trapped between the two plungers and in consequence they move as one solid plunger. This gives the idling-valve assembly a high momentary rate, such as is suitable to maintain steady idling.

For slow movements and permanent changes of position, the fuel-oil pressure on either side of the orifice can equalise themselves, due to the steady flow of oil through the orifice. In this case, since there is no hydraulic unbalance, the inner plunger (N) will take up a position which depends not only on the position of the outer plunger (P), but also on the rate of the inner and outer springs (M) and (O), since their loads must be equalised.

The outer spring (O) has one-half the rate of the inner spring (M). Thus a permanent change of position of the outer plunger (P) will cause the inner plunger (N) to change its position by one-third of the movement of (P).

The effect of this is to reduce the effective permanent rate of the governor, so that for any change of load or resistance at idling, the resultant change of idling rev/min is small.

Returning to the flow through the idling valve, it will be seen that a movement of the control rod and idling-valve plunger towards the 'open' position will allow the idling valve to move outwards, so that the inner plunger (N) closes off the slots. This restricts the flow of fuel oil through the valve, raises the 'closing' pressure and brings the servo piston and control rod back towards the 'closed' position. Thus the valve tends to hold the control rod, and also the engine speed, at a steady position depending upon the adjustment.

The actual speed of the engine, from idling to top speed, is selected by the position of the control pawl (U), as determined by the accelerator pedal position and linkage. The governor will adjust the fuel supply to bring the engine speed to the selected value, subject, of course, to the limitation imposed by the maximum power available.

The idling valve adjusting screw (Q), referred to as being carried midway on the governor lever, is used to adjust the sensitivity of the governor at idling.

What is the operation of the hydraulic governor fitted to the DPA distributor-type injection pump?

The hydraulic governor is of simple design, the whole being housed in a small casting attached to the side of the pump body. Its operation may be followed by reference to *Fig. 40*. The

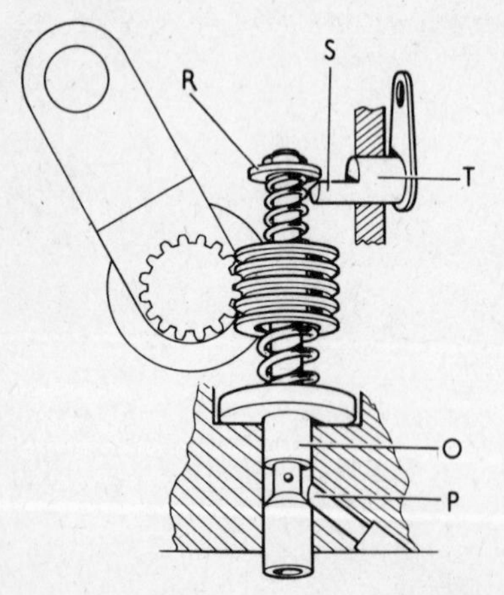

O. Metering valve
P. Metering port
R. Shut-off washer
S. Shut-off cam
T. Shut-off shaft

Fig. 40. Hydraulic governor mechanism of the CAV DPA distributor-type fuel-injection pump

metering valve (O) slides in a transverse bore in the hydraulic head, and an annular groove controls the passage of fuel to the metering port (P). The valve is hollow and this permits fuel at transfer pressure to reach the port through a cross-drilling. The valve stem carries the main governor spring and a light-idling spring, the latter being retained by a washer (R). The governor spring is loaded by the control lever by pinion and rack mechanism as shown. The larger disc on the valve stem acts as a damper to reduce sudden movements of the valve.

An overriding shut-off device is incorporated, the shaft (T) having a cam (S) cut at its end, so arranged that when the shut-off lever is operated, the cam lifts the washer (R) and lifts the metering valve, preventing any admission of fuel.

In normal operation, the metering valve is kept in position by a balance between the governor-spring pressure and the pressure of the fuel on its underside. An increase of engine speed raises the transfer pressure, and moves the valve against the spring until a balance is reached. When the control is moved towards the idling stop, the idling spring comes into play, the main-spring pressure being reduced until equilibrium is reached.

What is a smoke limiter or boost control?

An extra device, called a smoke limiter or a boost control, is normally incorporated on turbocharged engines to prevent excessive smoke emission.

Why is a smoke limiter or boost control needed?

When the accelerator pedal is depressed on a turbocharged engine, there is a slight delay before the turbine reacts to the exhaust gases and spins faster to boost the air delivered to the engine. Therefore, during acceleration, it is necessary to limit the amount of fuel delivered to match the amount of boost. If this is not done, a lot of black smoke will appear during the initial stage of acceleration.

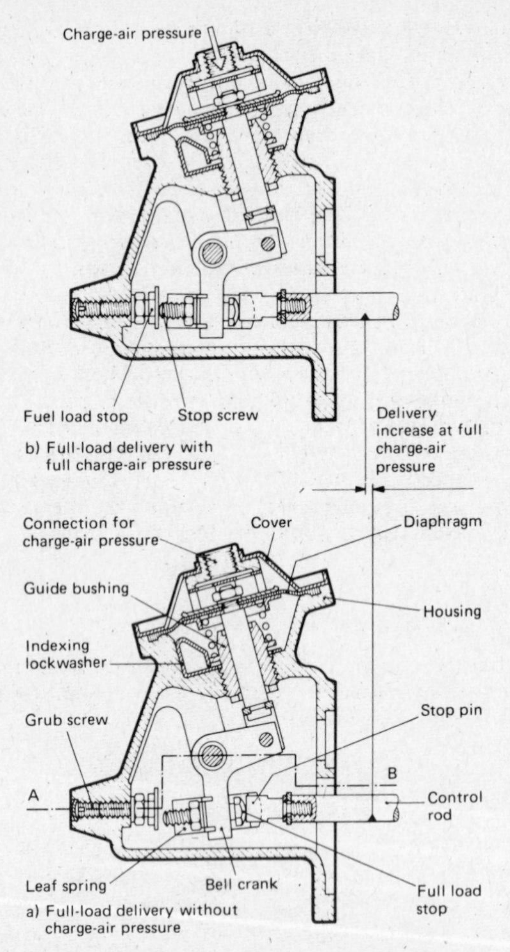

Fig. 41. Smoke limiter valve is fitted to the governor to limit the amount of fuel delivered during the initial phase of acceleration

How does this device work?

It consists of a diaphragm on a rod connected to a bell-crank whose other end is attached to the end of the main control rod in the injection pump. Normally the diaphragm is held in the raised position by a spring while air from the inlet manifold is passed through a hose to the chamber above the diaphragm. When there is no boost in the inlet manifold, the spring holds the diaphragm up so that the stop on the bell-crank lever prevents the full movement of the control rod. Therefore, less fuel than normal is delivered.

However, as the boost of the turbocharger increases, so the air pressure overcomes spring pressure and gradually moves the diaphragm and rod downwards. This pulls the stop away from the end of the control rod, so that more fuel can be delivered to the engine. At full boost, full fuel delivery can take place.

5

FUEL INJECTORS

What is the usual type of fuel injector?

The fuel injector is a spring-controlled valve inserted into the engine cylinder head (or in the side of the cylinder of opposed-piston engines) and allows the fuel, under pressure from the injection pump, to enter the cylinder or head in the form of a fine spray or mist. The injector is held in the correct position in the cylinder head by means of an injector holder.

How does a typical fuel injector work?

A complete fuel injector, which is also known either as a fuel valve, atomiser, nozzle or sprayer, consists of two parts, the injector valve and the injector body.

The injector valve takes the form of a barrel which, after being case-hardened, is ground and lapped to fit the injector body to the finest possible limits within which it will work freely. On one end of the valve a stalk is provided, while at the other end it is reduced in diameter to produce a stem upon which a valve face is formed.

Fuel is fed to the mouth of the injector through small tunnels bored vertically in the injector body which terminate in an annular reservoir or 'gallery' just above the valve seat (see *Fig. 43*). When the injector valve is raised from its seat in the injector body by the pressure of fuel being fed from the injection pump, the accumulated fuel in the gallery is pressed with great force through the hole or holes in the injector, thus forming a spray in the engine combustion chamber.

What is the construction of a typical fuel-injector holder?

The injector holder carries the valve spring and spindle against which the injector valve opens as a result of the fuel delivery pressure. At the lower end of the holder is a highly ground face which forms a joint with the flange of the injector body when tightened by means of the injector cap nut.

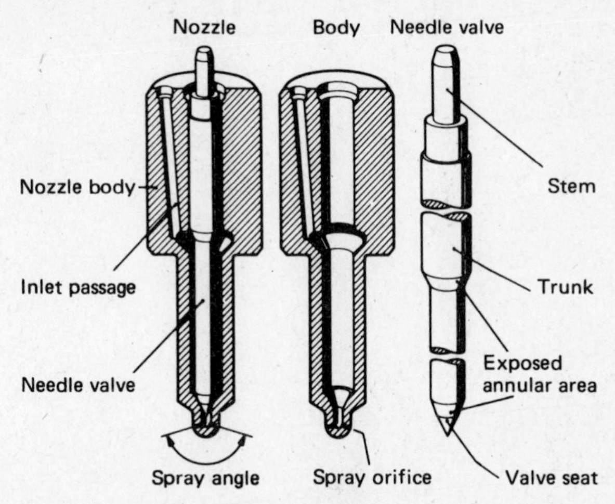

Fig. 42: Bosch fuel injector. A hole-type nozzle with long shank and long collar

Fuel is fed from the fuel-inlet connection through a boring in the injector holder which terminates in an annular semi-circular groove at the ground face of the flange of the injector body.

The slight leakage of fuel which accumulates within the injector holder, which is provided to lubricate the injector, is led away by a pipe connected to the leak-off connection.

What are the usual types of injectors?

The single-hole, multi-hole, long-stem, pintle, delay and Pintaux injectors.

Single-hole injector: The single-hole injector (*Fig. 43(A)*) has one hole drilled centrally through its body which is closed by the injector valve. In the CAV design, the hole can be of any

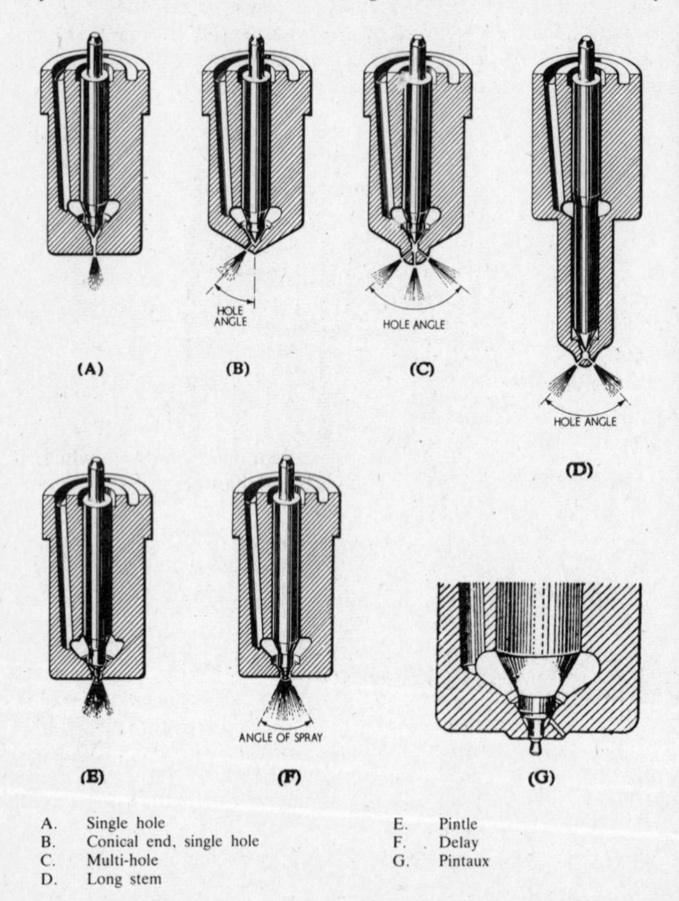

A. Single hole
B. Conical end, single hole
C. Multi-hole
D. Long stem
E. Pintle
F. Delay
G. Pintaux

Fig. 43. Types of fuel injectors

diameter from 0.2 mm upwards. A variation of this type is known as the conical-end injector (*Fig. 43(B)*).

Multi-hole injector: Multi-hole injectors (*Fig. 43(C)*) can have a varying number of holes drilled in the bulbous end under the valve seating, their actual number, size and disposition depending upon the requirements of the engine concerned.

Long-stem injector: For direct-injection engines where, owing to limited space between the valves in the cylinder head, it is not possible to provide adequate cooling for the standard short-stem injector, an alternative form of injector with a small-diameter extension has been developed. This is known as the long-stem injector and has an extended body, in the tip of which is provided the usual valve seating and dome for the injection holes. The valve stem is similarly elongated, but is a clearance fit in the body, the lapped portion of the barrel being confined to the section located above the fuel gallery (*Fig. 43(D)*). Thus, not only is the lapped guide raised to a higher level in the cylinder head, where there is usually adequate cooling, but, by virtue of the smaller diameter, it is also possible to provide cooling at the lower end.

Pintle injector: In the pintle injector (*Fig. 43(E)*), which is designed for use in engine combustion chambers of the air cell, swirl chamber or pre-combustion type, the valve stem is extended to form a pin or pintle which protrudes through the mouth of the injector body. Modification of the size and shape of this pintle can provide sprays varying from a hollow parallel-sided pencil form up to a hollow cone with an angle of 60° or more.

Delay injector: Certain pre-combustion chamber type engines, while requiring a pintle injector, demand different spray characteristics in order to obtain quieter running at idling speeds. This is accomplished by modifying the design of the pintle so that the rate of injection is reduced at the beginning of the delivery, the result being to reduce the amount of fuel in the combustion chamber, when combustion begins, thus diminishing 'diesel knock'. The modified injector is referred to as a 'delay injector' (see *Fig. 43(F)*). It should be noted, however, that this type of injector does not necessarily improve idling in every

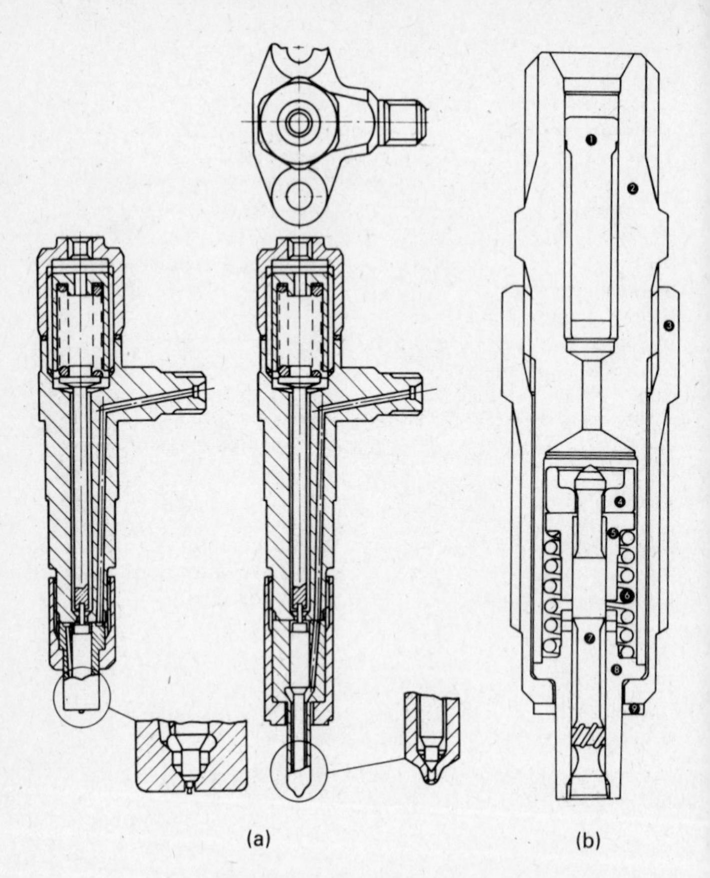

Fig. 44. (a) Lucas CAV pintle and multi-hole injectors. (b) Lucas CAV 'Microjector' screw-in pintle injector for car diesels. Holder body assembly: 1. Edge filter, 2. Nozzle-holder body, 3. Nozzle-holder capnut. Nozzle valve assembly: 4, Collar, 5. Lift stop, 6. Spring, 7. Nozzle valve, 8. Nozzle body, 9. Cylinder head sealing washer (Note: the 'Microjector' is actually much smaller than the other two.)

pre-combustion chamber engine, and should be selected only after a prolonged test embodying other considerations.

Pintaux injector: The Pintaux injector is a development of the pintle type, having an auxiliary spray-hole to assist easy starting under cold conditions (*Fig. 43(G)*). At engine-starting speeds, the needle valve is not lifted sufficiently to clear the pintle hole and the fuel is discharged through the auxiliary hole. At normal running speeds, however, when pressures in the fuel system are higher, the needle valve is withdrawn from the pintle hole, allowing the bulk of the fuel to be discharged through it.

How have injectors been changed in recent years?

Improvements in manufacturing processes have led to greater accuracy in the critical dimensions, but there have been a number of developments aimed at overcoming some problems.

What were these problems?

They were generally related to environmental legislation on smoke and noise emissions. Efforts have also been made to simplify installations in small engines for cars.

What changes have been made to reduce smoke emissions?

The one main change has been to reduce what is called the 'sac volume', this being the volume of fuel remaining in the nose of the injector body around the valve. It was found that this fuel had a tendency to leak out into the cylinder, thus increasing smoke emission. In addition, this leakage, however small, tended to create a build-up of carbon in the holes, and this led to a further deterioration in performance.

Another development aimed at improving cooling on DI engines has been the introduction of slim injector bodies. As a result of this improved cooling, better control of smoke emissions is achieved. Normally, injector bodies are around 25 mm diameter but the slimmer injectors are in the 18-20 mm diameter range, and thus can be fitted into less space while maintaining good cooling.

What other new developments have been made?

CAV has introduced what are called 'low-mass' injectors which have slim and short pushrods for the valves. Therefore, less stiff springs can be used. The result of this change is that the valve has less tendency to bounce on its seat, so the cut-off at the end of the injection is more abrupt. This not only reduces smoke levels, but can also reduce the noise level, since the injector makes a characteristic noise every time it strikes the seat. Therefore, if the force with which it strikes the seat can be reduced, as with the low-mass injector, then the noise level is reduced.

What changes have been made to suit car diesels?

To simplify installation, the injector holder is threaded on its periphery, so that it can be screwed into the cylinder head instead of being retained by studs and screws. In addition, Lucas CAV developed the 'Microjector', whose external dimensions are much smaller than in previous designs. The construction is quite different from that of the pintle injector; see *Fig. 44*.

If the engine is working irregularly and a faulty injector is suspected, how would you ascertain which cylinder has the faulty injector?

With the engine running irregularly, release the fuel-pipe union nut on each injector holder in turn until the offending injector is found. Releasing the union nut will prevent the fuel being pumped through the injector to the engine cylinder. Thus, the faulty injector will be the one which, when its fuel supply is cut off, gives no effect upon the engine running. When any of the other injectors, which are working correctly, are 'cut-out', the engine will run more unevenly.

How would you test a doubtful injector on its own piping?

First remove the securing nuts and withdraw the complete unit (injector holder and injector) from the cylinder head, turning it round the fuel-feed pipe so that the injector is pointing outwards, away from the engine. Next slacken the unions of the other injector-holder fuel-feed pipes (to prevent fuel being sprayed into their cylinders). Then turn the engine until the suspected injector sprays into the air, when it will be seen at once if the spray is in order.

The spray emitted should be quite symmetrical in form and finely atomised. Moreover, the valve should give a distinct grunt or buzz if the injector is in proper condition. If the injector emits an irregular or one-sided stream (in single-hole injectors) this is a sign of dirt on the seating. Similarly, if the individual sprays from a multi-hole injector are irregular, then the holes may be partly choked or there may be dirt on the seatings.

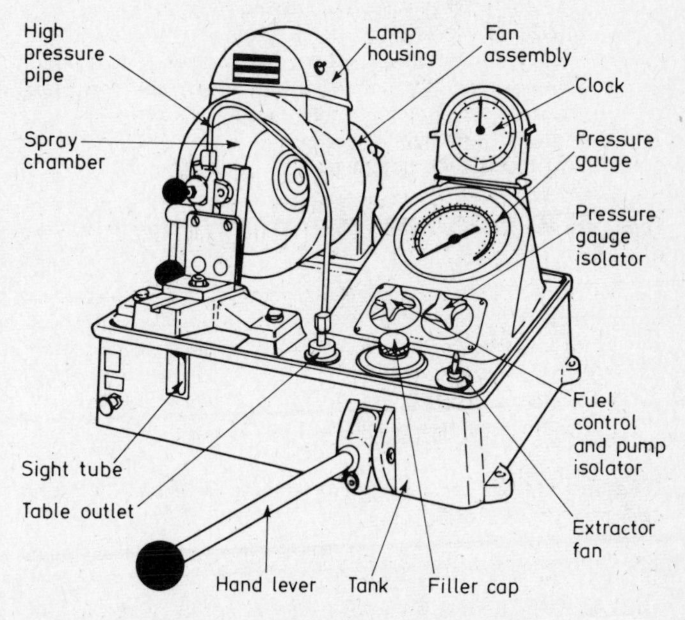

Fig. 45. Hartridge Testmaster in use to test injectors

What precaution must be taken when testing a fuel injector?

Great care should be taken to prevent the hands from getting into contact with the spray, as the working pressure will cause fuel to penetrate the skin with ease.

What faults may be found with fuel injectors?

(1) Incorrect operating pressure.
(2) Distorted spray form.
(3) Dripping injector.
(4) Injector valve not buzzing while injecting.
(5) Dirt between the injector valve and its seating.
(6) Cracked injector body.
(7) Broken injector-valve compression spring.
(8) Injector valve sticking in its guide.
(9) Compression-control spring not properly tensioned. (This is generally caused by the compression adjusting screw in the injector holder having slackened).
(10) A blued injector.
(11) Too much fuel escaping at the leak-off pipe.

What are the four possible causes of high injector pressure, and their remedies?

(1) Compression-spring adjusting screw shifted – adjust screw to the correct pressure.

(2) Injector valve seized up, corroded – renew injector valve and body.

(3) Injector valve seized up, dirty, sticky – clean the injector.

(4) Injector openings clogged with dirt – clean the injector.

What are the two most likely causes of low injector pressure and their remedies?

(1) Compression-spring adjusting-screw slackened – adjust screw to the correct pressure.

(2) Injector spring broken – replace spring and readjust pressure.

What can cause the form of the spray to become distorted?

An excessive carbon deposit on the tip, partially blocked injection holes or a damaged injector valve will distort the form of spray.

If the injector has a carbon deposit it should be cleaned and if the trouble is caused by a damaged injector valve the injector body and valve should be replaced.

What can cause a dripping injector?

A carbon deposit or a sticking injector valve can cause the injector to become leaky.

The remedy is to clean the injector valve. If this does not clear the fault, replace the injector valve and body.

What is the cause of an injector valve not buzzing while injecting?

This may be caused by the injector valve being too tight or binding, a leaky valve seat, or by a distorted injector cap nut.

The remedy is to clean the injector and examine the nut. If necessary, renew the injector body, valve and nut.

What is the object of the feeling pin fitted on some injector holders?

The feeling pin provides a means of checking whether the compression spring in the injector holder has broken or the injector valve is sticking in its guide, without removing the complete injector unit from the engine.

If you place a finger on the feeling pin when the engine is running, you should feel the pin giving a sharp kicking or pulsating movement. If there is no movement or only a feeble motion, this indicates either a broken spring or a stuck injector valve.

What can cause a blued injector?

Injector blueing may be caused by faulty installing, bad tightening or poor cooling.

The remedy is to renew the injector valve and body and check the cooling system.

What are possible causes of excessive leak-off of fuel at the leak-off connection, and their remedies?

(1) Injector valve slack – replace injector valve and body.

(2) Injector cap nut not tight – tighten nut.

(3) Foreign matter present between contact faces of injector and injector holder – clean the contact faces.

6

FUEL FILTERS

What types of elements are used in main fuel filters?

The most common for fuel-oil filter elements are wire gauze, cotton cloth, felt and special types of impregnated paper. Wire gauze will hold back only the largest particles and is practically useless for dealing with particles below 40 microns. Cotton cloth has been widely employed but this will filter only those particles larger than about 25 microns; felt is slightly more efficient and will hold back particles above 17 microns.

Specially-treated paper has proved, after severe tests, to be the most efficient, and it will filter much smaller particles than will either cloth or felt. Therefore treated papers are normally used nowadays.

What is the construction of a paper-element fuel-oil filter?

The CAV filter shown in *Fig. 46* makes use of strips of special filter paper spirally wound in V form. In addition to providing a better filtering media, the paper element gives an effective area several times greater than in other types.

The filter is of the cross-flow type. Fuel enters the filter through connection (4), passes down outside the element container, then up through the element and so out through the outlet connection (10). Dirty fuel is excluded from the clean side of the element by means of oil seals (5) at top and bottom of the element core, oil tightness of the seals being maintained by the pressure of spring (7). Paper elements must be replaced when choked as it is impracticable to clean them.

How do you renew the element of a main fuel-oil paper-filter?

Paper-element fuel-oil filters are not intended to be cleaned and the element must be discarded when choked, or at the mileage or running hours of the engine stipulated by the manufacturers.

To renew the element of a fuel-oil paper filter, first remove exterior dirt from the filter housing and cover and proceed as follows.

CAV F-type filter.–Drain the filter housing by unscrewing drain plug (8), *Fig. 46*, and slackening air-vent plug (2). Retighten the drain plug after the fuel oil has drained away.

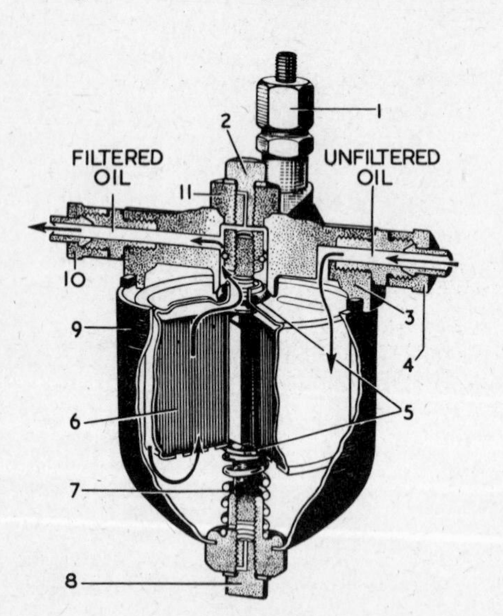

1. Gravity-vent valve
2. Air-vent plug
3. Cover
4. Fuel inlet connection
5. Oil seals
6. Paper element
7. Pressure spring
8. Drain plug
9. Filter level
10. Outlet connection
11. Cap nut

Fig. 46. Cut-away view of CAV F-type paper-element fuel-oil filter

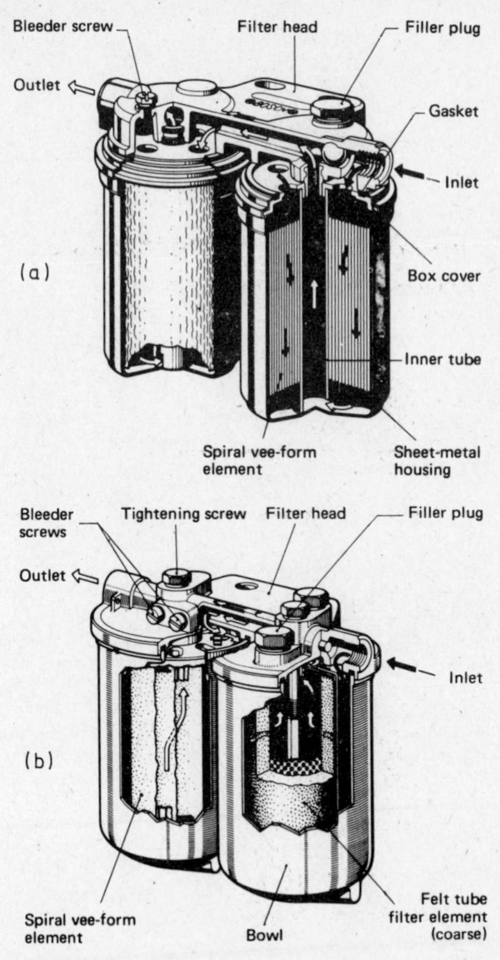

Fig. 47. Bosch fuel-oil filters: (a) a two-stage box filter, and (b) two-stage filter with replaceable filter elements

Unscrew cap nut (11), remove the filter housing and withdraw and discard the filter element.

Clean the housing with paraffin, fuel oil or petrol and blow dry with compressed air if available.

Check the condition of the element bottom sealing ring (5) and the gasket fitted between the filter housing and cover; renew if necessary.

Place the new upper sealing ring (5), supplied with the new element, in position and install the element in the housing.

Replace the housing, tighten cap nut and air vent the filter.

When are preliminary fuel-oil filters used?

In conditions under which the fuel is likely to be particularly badly contaminated, it is desirable to incorporate preliminary filters. For this duty, cloth, felt, combined cloth and felt-element type filters are used as well as water-trap pre-filters fitted with a comparatively coarse element. Where pre-filters are used, they are normally incorporated in the fuel line between the fuel tank and fuel-feed pump.

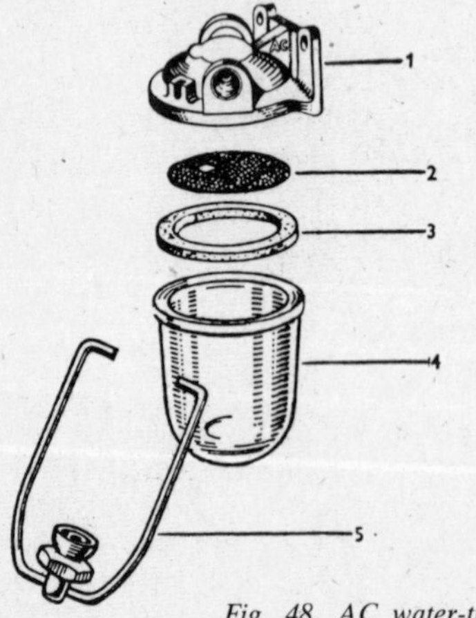

This type of pre-filter is usually connected in the fuel line between the fuel tank and feed pump. Its main purpose is to trap any water droplets in the fuel and so lessen the risk of corrosion troubles in the fuel-injection equipment.

The bowl, gasket and gauze screen can be removed and cleaned of sediment after slackening the bail-clamp nut.

1. Filter head casting
2. Gauze screen
3. Cork gasket
4. Glass bowl
5. Bail clamp

Fig. 48. AC water-trap fuel-oil filter

How do you clean a glass-bowl, water trap, preliminary fuel-oil filter?

To clean the AC bowl-type pre-filter shown in *Fig. 48* first remove exterior dirt from the bowl and head casting. Then slacken the bail nut and remove the bowl, gasket and gauze screen. Clean these parts with paraffin or fuel oil and allow to dry.

Inspect gauze screen and gasket for damage, and glass bowl for cracks; renew parts if necessary.

Assemble the gauze screen to the head casting, replace the gasket and bowl and tighten bail nut. Air vent the fuel-feed system.

Where are other filters sometimes used in fuel-system layouts?

A number of fuel-feed pumps are fitted with a gauze strainer, certain types of jerk-injection pumps have built-in felt pack filters installed in the pump housing, the DPA distributor-type injection pump has a nylon filter (gauze on early models) carried in the end-plate housing, and some injector holders incorporate an edge-type filter.

These filters should be periodically cleaned or renewed at the times specified by the manufacturers, although the jerk injection-pump final filter will not normally require attention until a complete overhaul of the pump is called for.

What is meant by 'venting' the fuel system?

It is necessary to vent or bleed the system, either if the fuel supply runs dry, or whenever the fuel-pipe lines have been disconnected, as is necessary when cleaning or renewing the filter element or overhauling a unit of the fuel system. This is to ensure that all air is removed from the system before attempting to start the engine.

What is the venting procedure with in-line jerk-injection pumps?

When bleeding the fuel system becomes necessary, first ensure that there is an adequate supply of fuel in the tank and then proceed as follows.

Where a vent plug is fitted in the fuel filter, this unit should be bled first. Slacken the vent plug and operate the hand-priming lever of the fuel-feed pump until bubble-free fuel is discharged from the vent. Tighten the vent plug as fuel is being discharged.

Next, bleed the fuel-injection pump by slackening the vent plug(s) on the pump and again operate the hand-priming lever on the fuel-feed pump until bubble-free fuel flows freely from the vent. Tighten the vent plug(s) while fuel is being discharged.

Thirdly, bleed the high-pressure pipes in sequence. With the engine idling, slacken either the vent screw (if fitted) in the nozzle holder or the union nut at the nozzle holder end of the pipes sufficiently to allow the fuel to seep out. Retighten the vent screw or union nut as soon as the fuel flows free from air bubbles.

It is also important to vent the system of an hydraulic governor. Operation of the hand primer on the fuel-feed pump, with the two vent plugs on top of the governor housing loosened (see *Fig. 39*) will expel any air from the circuit.

What is the venting procedure with the DPA distributor-type fuel-injection pump?

This question is answered in two parts. The first part gives the full priming (bleeding) procedure of the fuel system required after initial installation, or if the fuel system has become completely emptied. The second part gives the priming procedure necessary only when either or both the paper element of the main filter has been renewed or the filter bowl of a pre-filter (where fitted) has been cleaned.

General priming procedure: Slacken the vent (A), *Fig. 49,* on the front side of governor-control cover (mechanical governor) or the top of control-gear housing (hydraulic governor). Slacken

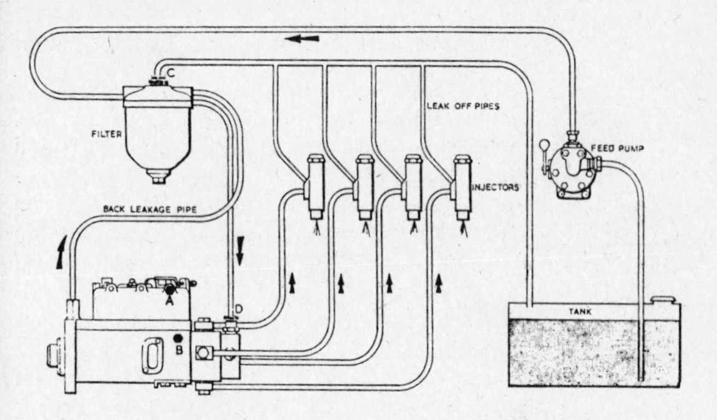

Fig. 49. Typical fuel system for DPA distributor-type injection pump with fuel-feed pump. Letters (A), (B), (C) and (D) indicate air venting or bleeding points

one of the two hydraulic-head locking screws (B). If the pump is installed in such a position that one head-locking screw is higher than the other, the screw at the higher level must be slackened. In later type pumps, a vent valve is fitted to one of the head-locking screws.

Unscrew, by two or three turns, the vent plug (C) on the top of the filter cover (not return pipe to tank). Operate the priming lever on the fuel-feed pump, or the separate hand primer in a gravity-feed system, and when bubble-free fuel issues from each venting point, tighten the screws in the following order: (i) filter-cover vent screw (C); (ii) head-locking screw (B); (iii) governor-vent screw (A).

Slacken the pipe-union nut (D) at the pump inlet, operate priming device and retighten when fuel oil issues freely around the threads.

Slacken the unions at the injector ends of two of the high-presure pipes. Set the accelerator at the fully-open position and ensure that the 'stop' control is in the 'run' position. Turn the engine until bubble-free fuel oil issues from both injector pipes. Tighten the unions on the injector pipes, and the engine is ready for starting and running.

Priming procedure after changing filter element: With the vent plug (C), *Fig. 49,* on the filter cover removed, and the union at the filter end of the return pipe (filter to tank) slackened, operate the hand primer until fuel oil issues freely from filter-cover vent. Replace the vent plug, and continue to operate the hand primer until fuel oil issues freely from around the threads of the return-pipe union. Tighten the return-pipe union.

Slacken the union at the filter end of filter-to-injection-pump feed pipe, and operate the priming device until bubble-free fuel oil issues from around the union threads. Tighten the feed-pipe union. The pump and filter are now filled and primed and ready for further service.

7

ENVIRONMENTAL DEVELOPMENTS

What are the main changes that have been made to diesel engines recently?

Among the main changes have been:

(1) The exploitation of the diesel for cars.
(2) Reductions in smoke, noise and gaseous emissions.
(3) An increase in power outputs with the aid of turbocharging.

Do car diesels differ from those used in trucks?

Since the diesel engine has a lower performance than the petrol engine, it must not be too heavy when installed in a car, nor must it be too expensive. Therefore, car diesels tend to be based on petrol engines to avoid excessive weight and production costs. In contrast, since diesel engines used in vans and trucks are designed for long life and hard usage, they are heavier than equivalent petrol engines, but usually last much longer.

Are there any technical differences?

The main technical difference between car and commercial vehicle diesels is in the governor which for a car engine must have characteristics closer to those of a petrol engine than to those of a truck engine. Therefore, the shut-off of the fuel supply at maximum rated speed is much more gradual. All car diesels are IDI units, of course.

What has been done to reduce smoke, noise and gaseous emissions?

Many changes have been made, or are being made, to reduce emissions. First, over the past few years, manufacturing tolerances have been reduced substantially, so that the compression ratio can be controlled very closely. There has also been the introduction of low-sac volume injectors (see *Chapter 5*).

Are any major changes being introduced to reduce gaseous emissions?

Generally, with detail changes to the induction and exhaust systems, and with retarded injection timing, gaseous emissions can be reduced sufficiently.

Does retarded timing affect anything else?

It also reduces the noise level but can impair power output and fuel consumption. However, if fuel is injected at higher pressures than in the past, so that injection is completed more quickly, then no loss in performance need be incurred. It is for this reason that the trend is towards the use of higher injection pressures and shorter injection periods.

Have any new combustion systems been developed to combat emissions?

Yes. One of the most interesting is the Perkins Squish Lip (SL) system (*Fig. 50*), coming into use first on Perkins industrial engines. In this design the combustion chamber is a small flat-bottomed bowl in the piston. The walls slope inwards towards the top, so that the throat is relatively small. Since the chamber is smaller than normal, the compression ratio of 19:1 is correspondingly higher, whereas on DI engines this is normally in the range 14–16:1. The combustion chamber imparts very high-speed motion to the air as the piston approaches top dead centre, so giving a good match between air movement and fuel delivery at all engine speeds.

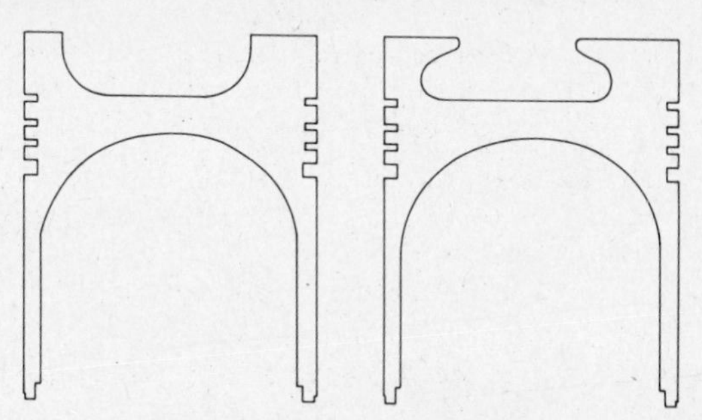

Fig. 50. Comparison of normal toroidal combustion chamber, left, and Perkins Squish Lip, right, which is actually much smaller since it operates with a compression ratio of about 19:1

As a result, this engine gives low gaseous emissions, it is quieter than normal engines, yet it can operate at higher speeds than other DI engines. Generally, DI engines do not operate well at speeds above about 3000 rev/min, but the SL unit is capable of operating at over 5000 rev/min on small engines.

One feature of the engine is that it can operate with injection timing much further retarded than in normal units, without any loss in power output or fuel consumption. Therefore, this SL system gives reduced noise levels because the combustion is quieter than normal.

How else can engine noise be reduced?

Apart from reducing combustion noise, the main means of reducing noise levels are:

Reduction of amplification of the noise.

Addition of shields so that the noise is not transferred.

What means are being used currently?

First, some reduction of combustion noise is being achieved by retarded injection and by turbocharging, the latter making combustion smoother and therefore quieter. Second, most engines have some form of shielding, although this may be attached to the vehicle chassis. Then, special rocker covers which damp vibrations are in use on many larger engines to reduce amplification of noise in this area. The same treatment is applied to timing covers and sumps in some cases to reduce noise further. Also, Lucas CAV has developed a range of injection pumps with reinforced housings to reduce noise.

In addition, many engines now have stiffened cylinder blocks which vibrate less and so generate less noise. Some engines can also be equipped with close-fitting shields to aid noise reduction, and usually these consist of a steel or aluminium shield separated from the engine casting by glass fibre or a plastics moulding.

A combination of these minor developments and these shields usually reduces the noise level of the engine so that on installation in the truck a 3 dBA reduction is obtained. This is on the ECE drive-by test, and means that the noise level is reduced from 87–89 to 84–86 dBA.

What other means are likely to be used in the future?

One promising avenue to improvement is to use pistons with reduced clearances in the cylinder bores. However, for reduced clearances to be successful, thermally-controlled pistons, possibly with steel inserts, are needed and these are under development.

An alternative is to use cast iron pistons with very thin walls. These need be no heavier than aluminium pistons, and have the advantage of greater resistance to thermal loads. They may be used, therefore, on very highly-rated turbocharged engines.

Can the engine structure be altered to reduce noise levels?

Since the main vibrations are amplified by the walls of the crankcase, a radical redesign of the cylinder block can reduce the noise level of the bare engine by over 10 dBA. In this structure, developed at Southampton University, the cylinder block resembles a frame, and this is clad on its sides with detachable covers made from a vibration damping material. Also, the main bearing caps are built up onto a beam to form a lower structure.

Less major changes, but with the careful use of covers and the repositioning of some assemblies, can also make a substantial difference to the noise level, as was shown with the Quiet Lorry project in which a Rolls-Royce 260 kW engine was used. The timing gears were placed at the back (immediately forward of the flywheel) where the crankshaft is much stiffer than at the front. With some redesign of the cylinder block, and many changes to the vehicle, the overall noise level on the drive-by test was reduced from about 89 to 81 dBA.

What changes are being made with turbochargers?

On big engines, the trend is to use higher turbocharging pressures, so that the engine can operate at lower speeds but give more power. Lower compression ratios are used to prevent the stresses becoming too high. The result is to improve the power/weight ratio of the vehicle. For example, for an output of 200 kW a naturally-aspirated engine of about 13 litres is needed. At the current rate of turbocharging, a 10-litre engine is needed, but with the high rate of turbocharging now possible, a 7.5-litre engine would be adequate. Such an engine would weigh about 700 kg, compared with about 1100 kg for the 13 litre unit.

What other developments have been introduced with turbochargers?

An important refinement with the turbocharger has been the addition of what is called a 'wastegate' (*Figs. 51* and *52*). This is a by-pass valve situated in the turbocharger (or as a separate unit) which allows some of the gases to by-pass the turbine.

The wastegate normally takes the form of a poppet valve, and in the neater designs this is mounted in the sidewall of the inlet passage to the turbine. Mounted on the stem of the valve is a diaphragm in a chamber connected by a hose to the intake manifold. Therefore, when the pressure in the inlet manifold reaches a preset pressure it exerts a sufficient force on the diaphragm to open the wastegate valve. As a result, some of the exhaust gases by-pass the turbine, and the inlet manifold pressure (the boost pressure) remains constant.

With this system, it is possible to use a smaller turbocharger which builds up the boost pressure much more quickly and at lower speeds. As a result, the maximum torque is increased more than normal, although the full potential power output of

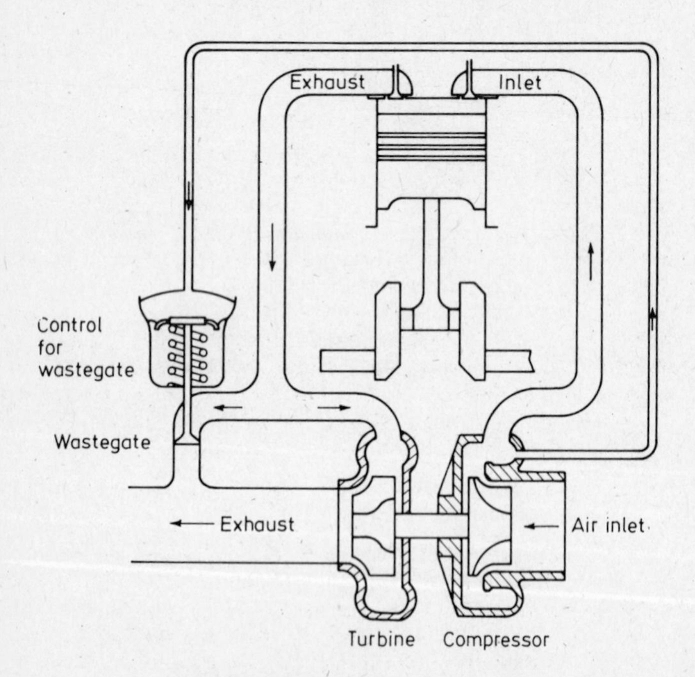

Fig. 51. A turbocharger and wastegate

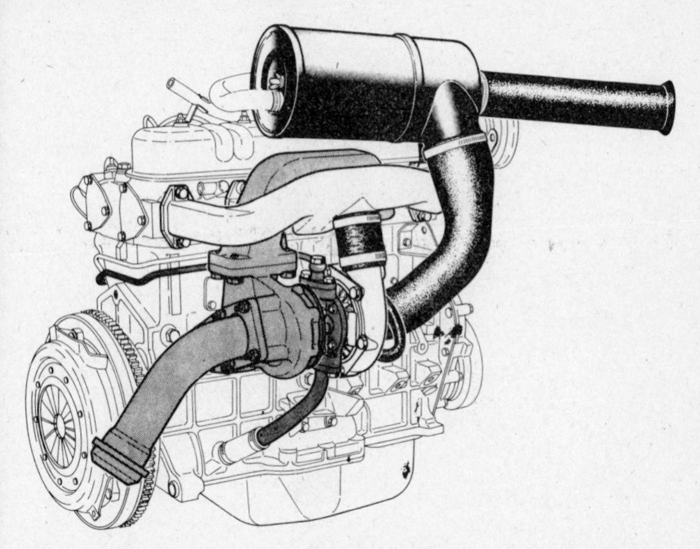

Fig. 52. The wastegate can be installed in the turbine housing to allow gas to by-pass the turbine when the pressure in the inlet manifold reaches a certain value

the turbocharger may not be realised. In practice, the slight loss in theoretical power output is insignificant, while the extra boost at low speeds is a decided advantage.

8

AUXILIARY EQUIPMENT

What auxiliary equipment is normally fitted to a diesel engine?

The auxiliary equipment depends on the application, but virtually all engines have oil and water pumps, though these are not really auxiliaries. The oil pump delivering lubricant to critical surfaces, such as bearings, cylinder walls and valve gear, is built into the engine and is essential. The water pump is not strictly essential, in that an engine can be cooled by thermo-syphoning, but for maximum efficiency, a pump to circulate the water under pressure is needed. Although it is usually bolted to a face of the cylinder block or head, the water pump is also incorporated into the engine in that it must be matched to the passages, and thus is really an integral part of the engine. Therefore, the main auxiliaries on automotive units are the starter motor, the alternator, the power steering pump and the compressor.

What types of starter motor are used on modern diesels?

Electric motors are almost universal, although some industrial engines have a 'spring' motor in which the operator winds up a spring whose energy is then released to rotate a shaft and crank the engine over. Generally, the electric motor is a dc unit with a pinion gear on its armature shaft. This pinion can be moved into mesh with a ring gear shrunk on to the flywheel of the engine.

What types of electric starter motor are used on diesels?

Two types are in use, the starter with a Bendix drive and the pre-engaged starter.

In the Bendix drive system the pinion is carried on a helix on the armature shaft, and it is held out of mesh by a light spring.

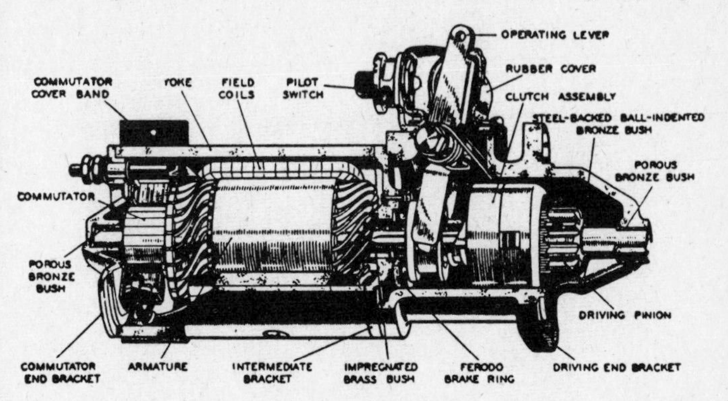

Fig. 53. Lucas M45G pre-engaged electric starter motor with manually-operated pinion

When the starter motor is rotated, the inertia moves the pinion along the helix so that it engages with the ring gear. Although this system is suitable for small engines, it is prone to disengage as soon as the engine fires, even if the engine 'dies' immediately. Therefore, it is not suitable for starting under difficult conditions, or in cold climates. For this reason the pre-engaged starter (*Fig. 53*) is now being used on virtually all diesels.

What is a pre-engaged starter?

In the pre-engaged starter there are two stages in the operation. In the first stage a solenoid moves the pinion into mesh with the ring gear. In the second stage the pinion is rotated, but only when it is in mesh – unlike the Bendix type in which, of course, the pinion is rotating before it engages the ring gear.

The solenoid can be a separate unit mounted on the body of the starter motor, in which case it actuates the pinion through a fork. Or, the solenoid can be mounted integrally with the starter, in which case it may be mounted co-axially with the armature shaft. One advantage of the type with separate solenoid is that if the pinion jams in the ring gear it can be freed manually. Also, the solenoid can be renewed easily, if necessary.

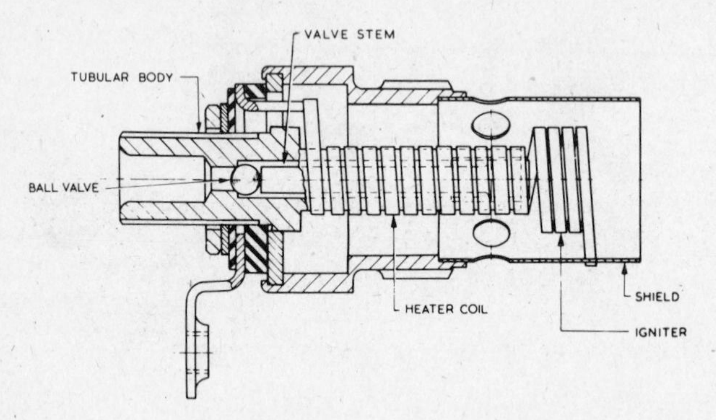

Fig. 54. CAV Thermostart type 537 cold-starting aid

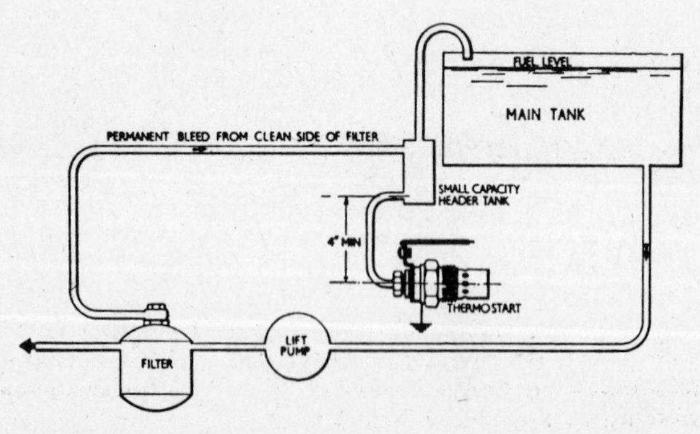

Fig. 55. Typical lay-out for CAV Thermostart type 537 cold-starting aid

Are any starting aids needed on diesels?

Yes, starting aids are needed on IDI and some DI engines to assist starting in cold weather. The fuel may not atomise well, and in the IDI swirl or pre-chamber the spontaneous combustion temperature may not be reached.

What sort of starting aids are used?

In IDI engines, 'glow plugs' are inserted into the pre-chamber. The glow plug is screwed into the cylinder head so that its nose projects into the pre-chamber. When the 'ignition' is switched on, electrical current flows to the plug (which is normally heated to about 800°C, although some new designs are heated to higher temperatures). Thus after about 20 seconds it is possible to start the engine, even if the ambient temperature is –20°C. At higher temperatures the delay is shorter.

What starting aids are used in DI engines?

DI engines start much better than IDI engines at low temperatures, so often no special starting aid is necessary. However, if a low compression ratio is used, or the vehicle is used in very cold areas, some method of heating the air before it enters the engine is needed.

Typical of these starting aids is the Lucas CAV 'Thermostart', which consists of an extra injector and igniter as it is injected, so that the air entering the engine is warmed sufficiently to start the engine. See *Figs. 54* and *55*.

As engines are becoming more highly turbocharged, lower compression ratios (under 14.5:1) are being used. Therefore starting aids are becoming more common on DI engines.

How is the alternator fitted to a diesel engine?

Normally, the diesel engine carries the alternator on brackets from the side of the cylinder block, the drive being by V belt – usually the same belt drives the water pump, fan and alternator.

Therefore, there is an adjusting strap between the engine and alternator so that the belt can be tensioned correctly, and correct tension is important. If the belt is too tight the bearings of the alternator will be overloaded, and if it is too slack the alternator will not charge the battery adequately, especially at low speeds. The method of adjustment depends on the application and the length of belt run between pulleys, but on smaller units correct tensioning of the belt should allow the longest run of the belt to be depressed by about 12 mm by hand halfway between the pulleys.

On some larger engines, the alternator is driven from the main timing gear train. The aim is to eliminate the need for belts, which are less reliable than gears, of course.

Why are alternators used?

DC generators were used for many years, but the ac alternator is now standard on virtually all engines. A rectifier is needed to produce dc current, but the big advantage of the alternator is that it generates electricity at much lower speeds than a dc generator, and can also operate safely at higher speeds. Initially, alternators with separate rectifiers were used, but now the trend is for the rectifier and voltage regulator to be built into the unit.

Why are compressors used?

The air braking system on an automotive unit needs a compressor and this should be able to compress air to about 700 kN/m^2 (100 lbf/in^2). Reciprocating compressors, which have been in use for many years, have either one or two cylinders, according to the air-flow needed, and in some cases the cylinder heads are water-cooled. The compressor is driven by the timing gears, and is often mounted in tandem with the fuel injection pump. To cut weight, noise and vibrations, various alternative types of compressor have been developed, the most promising being the Wankel type which is very small when compared to a reciprocating compressor. The efficiency of the latest designs is also high.

Are compressors used on all automotive diesels?

No. On some smaller units an exhauster is used, while on others the intake manifold is throttled to provide depression sufficient to actuate a conventional vacuum servo braking system. The exhauster, driven by the timing train, is usually a rotary pump of some kind – such as a vane pump – and it exhausts a reservoir for a vacuum braking system. It is common on smaller truck engines.

USEFUL UNITS AND METRIC CONVERSION FACTORS

Abbreviations

m = metre
g = gram
t = tonne (or metric ton)
N = newton (SI unit of force)
J = joule (SI unit of energy)
W = watt (SI unit of power)
s = second
M = mega (× 1 million)
k = kilo (× 1 thousand)
c = centi (1 hundredth)
m = milli (1 thousandth)
μ = micro (1 millionth)

Length

1 in = 25.4 mm
0.001 in = 0.0254 mm
1 mm = 0.03937 in
1 micron (μm) = 39.37 μin

Area

1 sq. in. (in^2) = 645.16 mm^2
= 6.4516 cm^2
1 cm^2 = 0.155 in^2

Volume

1 cu. in. (in^3) = 16.387 cm^3
1 UK gal. = 4.546 litres
1 cm^3 = 0.061 in^3
1 litre = 1000 cm^3 = 61 in^3

Mass

1 lb. = 0.4536 kg
1 ton = 1016 kg
1 kg = 2.205 lb
1 t = 1000 kg = 0.9842 ton

Force

1 lbf = 4.448 N
1 tonf = 9.964 kN
1 kgf = 9.807 N
1 N = 0.2248 lbf

Conversions between lbf, kgf, etc are as for mass units

Torque

1 pound-force foot (lbf ft)
= 0.1383 kgf m
= 1.356 N m
1 kgf m = 7.233 lbf ft
= 9.8067 N m
1 N m = 0.102 kgf m
= 0.7376 lbf ft

Pressure or stress

1 lbf/in^2 = 0.0703 kgf/cm^2
= 6.895 kN/m^2
1 tonf/in^2 = 1.575 kgf/mm^2
= 15.444 MN/m^2
1 kgf/cm^2 = 14.223 lbf/in^2
= 98.067 kN/m^2
1 N/m^2 = 0.000145 lbf/in^2
1 bar = 14.50377 lbf/in^2
= 10^5 N/m^2

Energy (work, heat)

1 ft lbf = 0.1383 kgf m = 1.356 J
1 Btu = 1.055 kJ
1 kJ = 102 kgf m = 737.6 ft lbf

Power

1 horsepower
(hp) = 550 ft lbf/s
= 1.0139 metric hp
= 76.04 kgf m/s
= 745.7 W
1 metric hp = 75 kgf m/s
= 735.5 W
1 ft lbf/s = 0.1383 kgf m/s
= 1.356 W
1 watt = 0.7376 ft lbf/s
= 0.102 kgf m/s
= 1 J/s = 1 N m/s
1 kW = 1.341 hp
= 1.36 metric hp

INDEX

QUESTIONS & ANSWERS ON

MOTOR CYCLES

G. FORSDYKE
Formerly Staff Writer 'Motor Cycle'

An easy to read book in question and answer format for the rider who wishes to know more about the machine he owns. It also covers the much wider aspects of the choice of machine for the novice and gives details of the capabilities and suitability of the various machines on the market.
An introduction to the sport in its many and varied aspects is given together with details of the growing youth sport movement. For the man interested in maintaining his own machine there are chapters giving details of servicing, necessary at various intervals, and the subject of machine safety is gone into at length.
Specific chapters dealing with aspects such as brakes, lighting, carburation, etc. go into detail to aid maintenance of these all-important items.
And for the man taking a machine onto the road for the first time, legal requirements including the need for licences, test certificates and insurance are included.

CONTENTS: Types of Machine; Engines; Carburation; Ignition Systems; Gearboxes; Motor Cycle Parts; Electrical Systems; Routine Maintenance; Licensing, Insurance, Legal Notes; On the Road.

128 pages **1976** **0 408 00232 8**